Lecture Notes in Electrical Engineering

Volume 755

The book series *Lecture Notes in Electrical Engineering* (LNEE) publishes the latest developments in Electrical Engineering - quickly, informally and in high quality. While original research reported in proceedings and monographs has traditionally formed the core of LNEE, we also encourage authors to submit books devoted to supporting student education and professional training in the various fields and applications areas of electrical engineering. The series cover classical and emerging topics concerning:

- Communication Engineering, Information Theory and Networks
- Electronics Engineering and Microelectronics
- Signal, Image and Speech Processing
- Wireless and Mobile Communication
- Circuits and Systems
- Energy Systems, Power Electronics and Electrical Machines
- Electro-optical Engineering
- Instrumentation Engineering
- Avionics Engineering
- Control Systems
- Internet-of-Things and Cybersecurity
- Biomedical Devices, MEMS and NEMS

For general information about this book series, comments or suggestions, please contact leontina.dicecco@springer.com.

To submit a proposal or request further information, please contact the Publishing Editor in your country:

China

Jasmine Dou, Editor (jasmine.dou@springer.com)

India, Japan, Rest of Asia

Swati Meherishi, Editorial Director (Swati.Meherishi@springer.com)

Southeast Asia, Australia, New Zealand

Ramesh Nath Premnath, Editor (ramesh.premnath@springernature.com)

USA, Canada:

Michael Luby, Senior Editor (michael.luby@springer.com)

All other Countries:

Leontina Di Cecco, Senior Editor (leontina.dicecco@springer.com)

**** This series is indexed by EI Compendex and Scopus databases. ****

More information about this series at http://www.springer.com/series/7818

Abhijit Biswas · Raghvendra Saxena · Debashis De
Editors

Microelectronics, Circuits and Systems

Select Proceedings of 7th International Conference on Micro2020

Springer

Editors
Abhijit Biswas
Institute of Radio Physics and Electronics
University of Calcutta
Kolkata, West Bengal, India

Raghvendra Saxena
Government of India
Defence Research and Development
Organisation
Delhi, India

Debashis De
Department of Computer Science
and Engineering
Maulana Abul Kalam Azad University
of Technology
Kolkata, West Bengal, India

ISSN 1876-1100 ISSN 1876-1119 (electronic)
Lecture Notes in Electrical Engineering
ISBN 978-981-16-1569-6 ISBN 978-981-16-1570-2 (eBook)
https://doi.org/10.1007/978-981-16-1570-2

This Springer imprint is published by the registered company Springer Nature Singapore Pte Ltd.
The registered company address is: 152 Beach Road, #21-01/04 Gateway East, Singapore 189721, Singapore

Contents

About the Editors

Abhijit Biswas received his B.Tech., M.Tech. and Ph.D. (Tech.) degrees from the University of Calcutta. In 1999, he joined the University of Calcutta as a lecturer in the Department of Radio Physics and Electronics where he is currently working as a Professor. He worked at Jadavpur University as a lecturer in Electronics Science during 1997–1999 and as a Reader in the Department of Electronics and Telecommunication Engineering during 2006–2008. His research interests include semiconductor physics, modeling, simulation, and characterization of electronic and optoelectronic devices including CMOSFETs, LEDs and LASER diodes, and circuits. He has published over 70 technical papers in peer-reviewed international journals.

Dr. Raghvendra Saxena received his B.E. degree in Electronics and Communication Engineering from G. B. Pant Engineering College, India, in 1997, M.Tech. degree in Microelectronics from IIT Bombay, in 2003, and a Ph.D. degree from IIT Delhi, in 2012. Since 1998, he has been associated with the Solid State Physics Laboratory, DRDO, Delhi, where he is currently heading a research group working on the development and testing of infrared detectors and their readout circuits. Dr. Sahai's research interests include infrared detectors, power electronic devices, and nanoscale VLSI devices & circuits. He has authored over ninety papers in various international refereed journals and conference proceedings. Dr. Saxena is the Editor of IETE Technical Review and SSPL's Technical Bulletin-Crystal. Dr. Saxena is an American Society for Quality (ASQ) Certified Reliability Engineer. He is also a Fellow of IEI, India, and IETE, India.

Prof. Debashis De earned his M.Tech. from the University of Calcutta in 2002 and a Ph.D. from Jadavpur University in 2005. He is a Professor and Director in the Department of Computer Science and Engineering, West Bengal University of Technology, India, and an Adjunct Research Fellow at the University of Western Australia, Australia. He is a senior member of the IEEE, a life member of CSI, and a member of the International Union of Radio Science. His research interests include mobile cloud computing and green mobile networks. He has published over 270 papers in peer-reviewed international journals and 200 international conference papers, six research monographs, and ten textbooks. He is an Associate Editor of IEEE ACCESS journal and Editor of the Hybrid Computational Intelligence journal.

Advanced MOSFETs

High Switching Performance of Novel Heterogeneous Gate Dielectric—Hetero-Material Based Junctionless-TFET

Samriti Sharma and Rishu Chaujar

1 Introduction

A conventional MOSFET faces a difficulty in attaining a low-thermal budget because of the necessity of complex fabrication and large doping-concentration gradient due to the existence of two source/channel and channel/drain junctions. Scaling down the MOSFET dimensions in accordance with the International Technology Roadmap for Semiconductors (ITRS) guidelines becomes very challenging due to the doping junctions present between the source/channel and channel/drain boundaries beyond 32-nm nodes technology [1]. In order to withstand the downscaling issues with time various novel structures have been evolved by the researchers. In past few years junctionless field-effect transistors (JLFET) have attracted the interest of many researchers due to the absence of doping junctions [2, 3]. Unlike MOSFET, the doping concentration of JLFET is the same all over the source, channel, and drain regions. Owing to the absence of concentration gradient in the lateral direction of the channel, this device is quite easier to fabricate along with improved electrical performance and better uniformity than MOSFET. The junctionless concept has been explored in many devices such as TFETs [4], FinFETs [5], Negative capacitance FETs [6], Nanowire FETs [7]. TFETs are the most promising candidates in present technology over MOSFETs owing to its steeper sub-threshold swing (<60 mV/decade) and lower leakage current, [8–10]. The Tunnel FETs work on the fundamental mechanism of band-to-band tunnelling [11, 12]. By applying the junctionless technology, the result characteristics of a TFET can be enhanced in the form of higher ON current, steeper sub-threshold swing along with fast current switching ratio [13]. Along with junctionless technology, the direct bandgap III-V compound semiconducting materials also aid to accomplish higher ON state current by incorporating lower bandgap and higher bandgap materials in the Source and channel region [14–16]. Further, enhancement

S. Sharma · R. Chaujar (✉)
Delhi Technological University, Rohini 110042, India
e-mail: chaujar.rishu@dtu.ac.in

A. Biswas et al. (eds.), *Microelectronics, Circuits and Systems*, Lecture Notes in Electrical Engineering 755, https://doi.org/10.1007/978-981-16-1570-2_1

in ON current and SS can be achieved by High-k gate oxide; however, it increases the leakage current and the ambipolar current in TFETs. So hetero-gate dielectric is better option in comparison to high-k dielectrics [17].

In this work, we have designed a hetero-gate-dielectric hetero-junctionless TFET (HD-HJLTFET), using III-V compound semiconducting materials. The InAs (lower bandgap) and GaAs (higher bandgap) are used in the source and channel regions, respectively, giving rise to a hetero-junctionless device, which provides better results as compared to Si-based TFET because of the higher energy bandgap and hence wider tunneling barrier in case of Si material-based devices. A heterogeneous gate dielectric engineering is executed in the proposed device, HD-HJLTFET by locating high-k and low-k material (SiO_2) in the oxide region under the control gate nearby the source side and drain side, respectively. The results are obtained for the optimization of the high-k stack material and stack length by using different dielectric materials in the stack. A comparison of our proposed device HD-HJLTFET is made with two conventional configurations namely Low-k HJLTFET (which uses only SiO_2 as the gate oxide under the control gate) and High-k HJLTFET (which uses only a high-k oxide under the control gate) is investigated in this paper for analog applications. The paper is structured in the following manner: Sect. 2 presents the device architecture and simulation, Sect. 3 comprises of results and discussion, and Sect. 4 the conclusion.

2 Device Architecture and Simulation

Figure 1a–d represents the graphic outlook of our entire device design framework [(a), (d) HD-HJLTFET, (b) Low-k HJLTFET, and (c) High-HJLTFET]. The doping concentration of all the aforementioned devices is uniform throughout the length with a charge concentration of 1×10^{19} cm^{-3}. The channel length of the device is 20 nm and the body thickness is 3 nm. The polar gate (PG) towards the source is used to modulate the source region polarity to p-type and the control gate (CG) modulates the current transport in the channel region. The work function of PG and CG is 5.93 eV and 4.7 eV, respectively. There is an isolation between PG and CG of 2 nm, which also works as the spacer between the two gates. In order to strengthen the gate controllability over the channel, dual gate technology is incorporated.

The results are evaluated after optimizing the high-k material as HfO_2 with optimized high-k length of 7 nm. The biasing conditions of the device are $V_{GS} = 1.5$ V and $V_{DS} = 1.5$ V for the ON state and $V_{GS} = 0.0$ V and $V_{DS} = 1.5$ V for the OFF state.

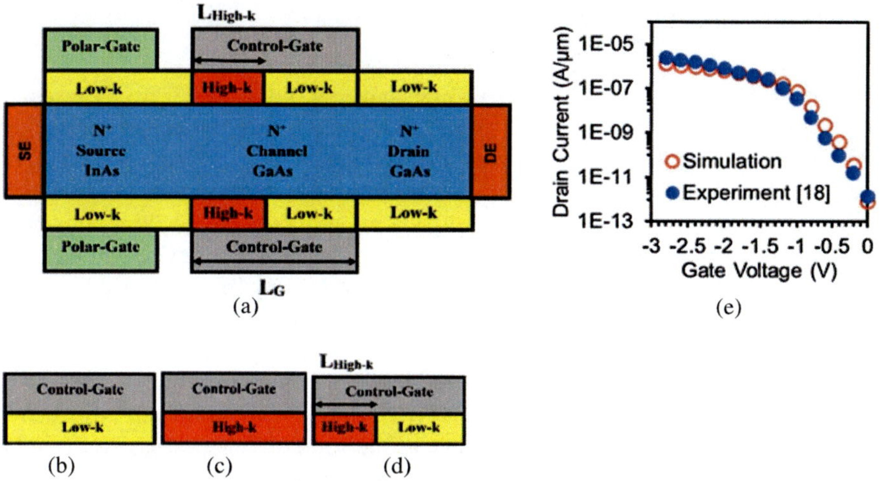

Fig. 1 Device architecture of **a** HD-HJLTFET, **b** Low-k HJLTFET, **c** High-k HJLTFET, **d** HD-HJLTFET. **e** Calibrated transfer characteristics

2.1 *Device Architecture and Simulation*

We have applied the non-local band-to-band tunneling model to calculate the excavation of the carriers across the source to channel boundary. We also invoked the bandgap narrowing and Auger recombination model along with SRH recombination model and Fermi Dirac Statistics to perform the intrinsic carrier calculations. To include the effects of parallel, perpendicular electric field mobility, temperature-dependent mobility, and concentration-dependent mobility, we invoked the CVT model. To include the quantum effects, we implemented the quantum confinement model of Hänsch et al. [18]. All simulations of our device are performed in Silvaco-ATLAS simulator [19].

2.2 *Calibration*

Before simulating our proposed device, we have validated the simulation setup with the experimental data extracted from the fabrication-based published work [20]. In order to do so, we deliberated the device parameters and the biasing conditions of our device exactly as that of the work reported in [20]. Figure 1e illustrates the validation of simulation setup with the experimental data in terms of the transfer characteristics.

The close proximity of the simulation and experimental results authorizes the model deliberations of the simulation setup.

3 Results and Discussion

The energy band diagrams of HD-HJLTFET and Low-k HJLTFET are represented in Fig. 2a for the OFF state, ($V_{GS} = 0.0\,\text{V}$, $V_{DS} = 1.5\,\text{V}$). The implementation of hetero-material InAs in the source and GaAs in the channel and drain, the bandgap difference between the two materials leads to a wider barrier width at the source to channel interface as evident from the figure. This results in minor flow of carriers due to the P-i-N diode leakage. As revealed in Fig. 2b, when the device is turned on, ($V_{GS} = 1.5\,\text{V}$, $V_{DS} = 1.5\,\text{V}$) by applying the gate biasing of 1.5 V at the control gate, due to the hetero-materials present at the source and channel, a local minimum is induced in the conduction band at the source to channel (S/C), boundary. Due to the application of high-k oxide towards the source, gate to channel coupling is strengthened and this local minimum induced in the conduction band edge bends further and overlaps the valence band edge of the source. This process elevates the tunneling probability of the carriers from the source to the channel because of the narrower tunneling barrier width at the S/C boundary as compared to the low-k oxide. Figure 2c illustrates the electron and hole concentration of HD-HJLTFET and Low-k HJLTFET, when no gate biasing is applied. This reveals that the devices attain the P-i-N doped configuration deprived of any physical doping being implanted. In

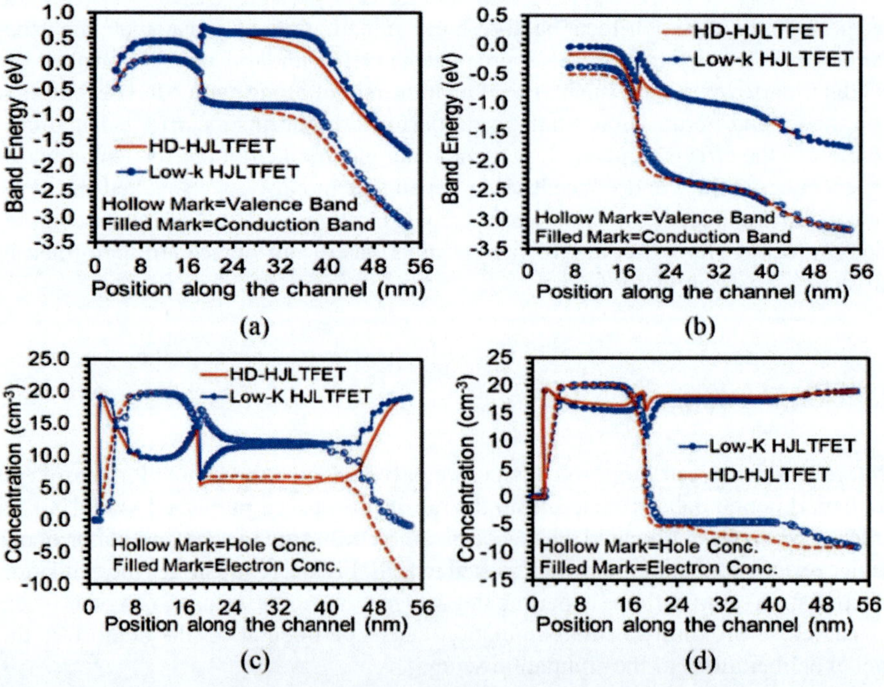

Fig. 2 **a** OFF state energy band diagram, **b** ON state energy band diagram, **c** OFF state carrier concentration, and **d** ON state carrier concentration of HD-HJLTFET and Low-k HJLTFET

the ONstate, as shown in Fig. 2d, When gate biasing is applied to CG, the electron concentration profile of the HD-HJLTFET increases under the control gate due to the high-k stack towards the source region. It gives rise to an n doped pocket region beneath the spacer region between CG and PG. Hereafter the barrier width gets shortened and causes to shift in the tunneling of the carriers towards the left of the junction. This fact is further supported by the results revealed by the non-local band-to-band electron tunneling rate of HD-HJLTFET Compared with Low-k HJLTFET as shown in Fig. 3a.

As depicted from Fig. 3a, on the application of high-k oxide close to the source, overlapping of conduction band local minimum edge and the valence band of the source leads to narrower barrier width as explained before. The electrons tunnel through the barrier with a higher tunneling rate in HD-HJLTFET as the electric field of CG is greater at the S/C boundary as compared to Low-k HJLTFET, as revealed in Fig. 3b, providing higher mobility to the carriers to tunnel through the junction whereas, low-k dielectric material diminishes the mobility of carriers due to lower gate to channel coupling. This improves the flow of the electrons across the hetero-junction and hence electrons start early tunneling at a higher rate as compared to Low-k HJLTFET. Among three devices as revealed in Fig. 3b, HD-HJLTFET and High-k HJLTFET show almost similar trend of electric field at the S/C intersection in contrary to Low-k HJLTFET. Whereas, the electric field of High-k HJLTFET is risen at the

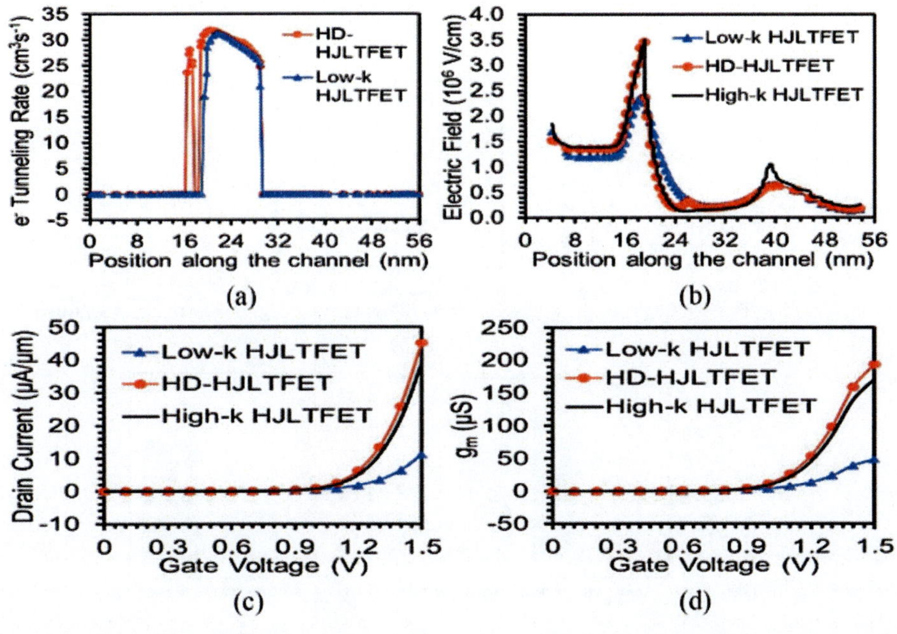

Fig. 3 **a** Non-local BTBT electron tunneling rate and **b** ON state electric field along the channel direction. **c** Transfer characteristics and **d** Transconductance, g_m of HD-HJLTFET compared with Low-k HJLTFET and High-k HJLTFET

D/C interface as compared to the other two devices as an outcome of high-k oxide at the drain side, which rises the drain to channel coupling. In Fig. 3c, a comparison of the transfer characteristics of HD-HJLTFET is displayed with Low-k HJLTFET and High-k HJLTFET. The ON current of HD-HJLTFET is 1.2 orders higher in magnitude than High-k HJLTFET and 4.04 orders higher in magnitude than Low-k HJLTFET. This increment in ON current is due to the channel modulation leading to larger tunneling probability of the electrons across the junction on the application of heterogeneous gate dielectric stack under CG. Another crucial parameter that determines the current driving efficacy of the device at a constant drain bias is the transconductance, g_m. To attain higher gain, the device should exhibit higher value of g_m. From Fig. 3d, it is observed that HD-HJLTFET shows superiority in terms of g_m in comparison to Low-k HJLTFET and High-k HJLTFET attributed to the enhanced driving current on the application of heterogeneous gate dielectric under CG.

In Fig. 4a it can be observed that the higher order transconductance coefficient, g_{m3} attains lowest value for HD-HJLTFET, which is 11.96% and 309.4% lower as compared to High-k HJLTFET and Low-k HJLTFET, respectively. The aforementioned reduction in g_{m3} for HD-HJLTFET supports improved linearity performance of the device because of lower amplitude. Figure 4b displays the comparison of device efficiency as a function of gate voltage, which is the ratio of transconductance to current. It is noticeable from Fig. 4b that hetero-dielectric device attains the highest device efficiency as compared to Low-k HJLTFET and High-k HJLTFET. The peak of device efficiency in HD-HJLTFET and Low-k HJLTFET is also attained

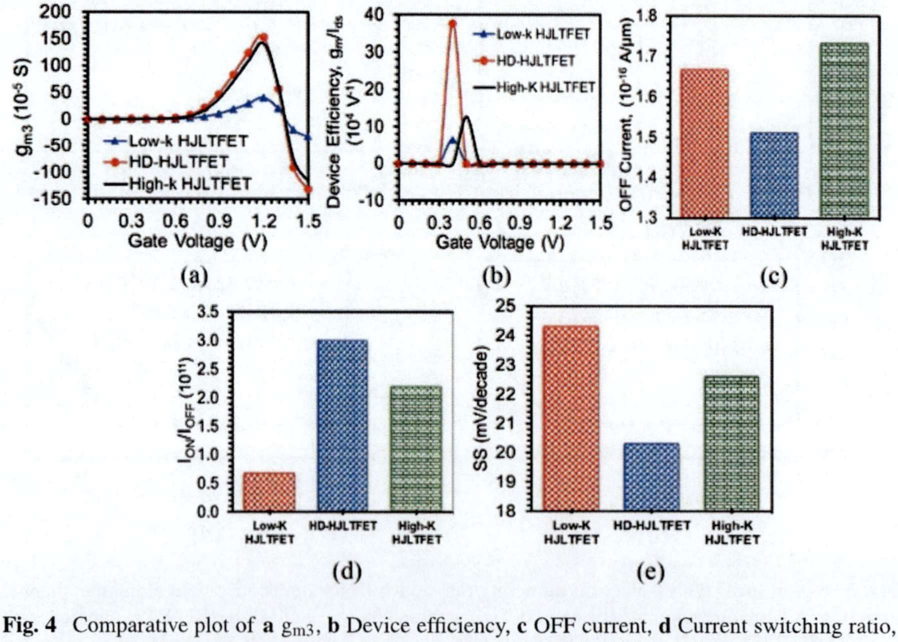

Fig. 4 Comparative plot of **a** g_{m3}, **b** Device efficiency, **c** OFF current, **d** Current switching ratio, I_{ON}/I_{OFF}, and **e** SS of HD-HJLTFET compared with Low-k HJLTFET and High-k HJLTFET

at lower gate voltage than High-k HJLTFET. Device efficiency of the proposed device is 194.5% and 480% higher than High-K HJLTFET and Low-k HJLTFET, respectively. In Fig. 4c, it is worth observing that a minimal leakage current (OFF current) is produced in HD-HJLTFET, which is 12.7% and 9.03% lower than High-k HJLTFET and Low-k HJLTFET, respectively. This is resulted because of the larger tunneling barrier width at the S/C interface when the device is in the OFF state. The High-k HJLTFET shows the highest value of leakage current in comparison to the other two devices, which is the result of higher electric field and drains to channel coupling at the D/C junction. Figure 4d illustrates the comparison plot of current switching ratio, I_{ON}/I_{OFF} of HD-HJLTFET, Low-k HJLTFET, and High-k HJLTFET. As clearly expressed by Fig. 4d, our proposed device, HD-HJLTFET exhibits the highest I_{ON}/I_{OFF} ratio, which is 1.36 and 4.5 orders higher in magnitude than High-k HJLTFET and Low-k HJLTFET, respectively. In Fig. 4e, our proposed device, HD-HJLTFET shows ~10% and ~16% smaller value of SS as compared to High-k HJLTFET and Low-k HJLTFET, respectively, which is a significant parameter for analog applications of TFET.

The optimization of dielectric material for the proposed device is investigated by employing different insulating materials, i.e., HfO_2 (k = 25), ZrO_2 (k = 22), Al_2O_3 (k = 9), Si_3N_4 (k = 7), and SiO_2 (k = 3.9), in the high-k region. In Fig. 5a, b, the transfer characteristics of HD-HJLTFET for different dielectric materials are illustrated. The results show a marked improvement in the form of ON current on increasing the dielectric constant of the high-k. HD-HJLTFET with HfO_2 (k = 25) shows the highest value of the ON current which is due to the higher gate coupling provided by the higher dielectric constant material. The physical oxide thickness is kept constant as 2 nm throughout the simulations for all the dielectric materials. Figure 5c displays the comparison plot of current switching ratio and SS as a function of different dielectric materials in the high-k region as mentioned before. We can observe that for Low-k HJLTFET device (SiO_2, k = 3.9), the I_{ON}/I_{OFF} is least among the chosen dielectric materials and as we increase the dielectric constant, k of the high-k region; the I_{ON}/I_{OFF} exhibits positive gradient for HD-HJLTFET. The reason

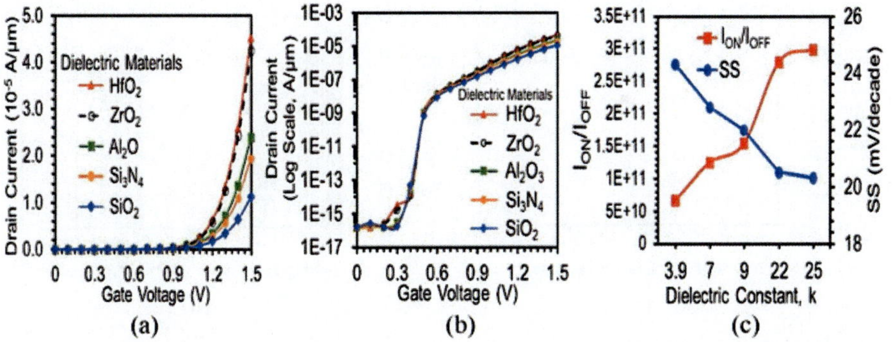

Fig. 5 Comparison of Transfer characteristics in **a** Linear scale, **b** Log scale for different dielectric materials, and **c** current switching ratio and SS variation for dielectric materials constants

behind such a trend is the increment in the electric field at the S/C intersection below CG owing to the existence of high-k region. The maximum I_{ON}/I_{OFF} is attained for HfO_2 in the range of 3×10^{11}, while I_{ON}/I_{OFF} of SiO_2, $k = 3.9$ (Low-k HJLTFET), shows the least value of 0.6×10^{11}. So, the optimized dielectric material for the high-k region is HfO_2. In order to optimize our proposed device, HD-HJLTFET, the length of the high-k region, $L_{High\text{-}k}$ has been optimized for high-k dielectric material HfO_2.

Figure 6a expresses the transfer characteristics of HD-HJLTFET as a function of gate voltage by varying the length of the high-k region, $L_{High\text{-}k}$. It can be observed that the ON current shows an increasing pattern as we reduce the $L_{High\text{-}k}$ from 20 to 7 nm, whereas the ON current again decreases for $L_{High\text{-}k} = 5$ nm. The $L_{High\text{-}k} = 20$ nm corresponds to High-k HJLTFET, where HfO_2 is located over the whole channel region. Figure 6b displays the OFF state current variation of HD-HJLTFET as a function of $L_{High\text{-}k}$. The OFF state current starts reducing as the $L_{High\text{-}k}$ is decreased from 20 to 10 nm and starts increasing for further reduction in $L_{High\text{-}k}$. The variation of current switching ratio, I_{ON}/I_{OFF} of HD-HJLTFET is presented as a function of $L_{High\text{-}k}$ in Fig. 6c, which shows an up-surging trend of I_{ON}/I_{OFF} ratio with an increment of ~36% as we decrease the $L_{High\text{-}k}$ from 20 to 7 nm. The I_{ON}/I_{OFF} ratio starts decreasing as the $L_{High\text{-}k}$ is further reduced to 5 nm. Figure 6d displays the comparison of SS of HD-HJLTFET as a function of $L_{High\text{-}k}$. It is clear from Fig. 6d that the optimized

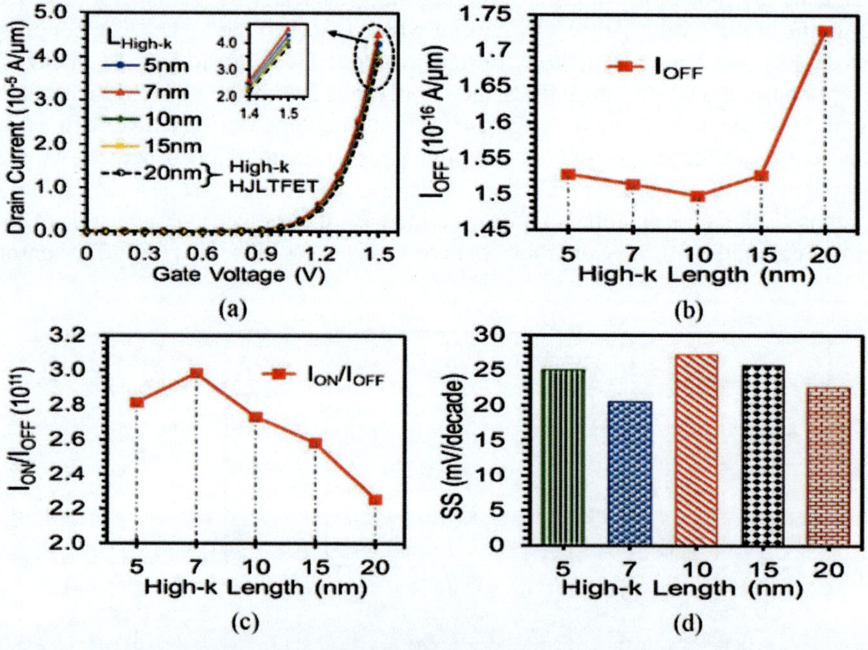

Fig. 6 Comparison of **a** Transfer characteristics, **b** OFF current, **c** Current switching ratio, I_{ON}/I_{OFF}, and **d** SS of HD-HJLTFET for different high-k gate stack lengths, L_{Stack}

Table 1 Result characteristics of HD-HJLTFET, Low-k HJLTFET and, High-k HJLTFET

	HD-HJLTFET	Low-k HJLTFET	High-k HJLTFET
I_{ON} (μA/μm)	45.2	11.2	39
I_{OFF} (A/μm)	1.5×10^{-16}	1.6×10^{-16}	1.7×10^{-16}
I_{ON}/I_{OFF}	3×10^{11}	0.6×10^{11}	2.2×10^{11}
SS (mV/decade)	20	24	22

$L_{High\text{-}k}$ is 7 nm, at which the SS is smallest. The optimization of $L_{High\text{-}k}$ depends upon the local minimum induced at the conduction band edge of the S/C interface. On reducing $L_{High\text{-}k}$, the conduction band well gets shallower resulting into minor tunneling of electrons. On the contrary, the increment in $L_{High\text{-}k}$ widens the conduction band well leading to poor transition between ON and OFF states. On comparing the results of current switching ratio and SS for $L_{High\text{-}k}$, the length of the high-k stack is optimized to 7 nm (Table 1).

4 Conclusion

In this work, we proposed junctionless TFET with a III-V compound semiconducting hetero-materials using a hetero-gate dielectric having a combination of high-k oxide at the source side and low-k oxide at the drain side. Our device, HD-HJLTFET shows marked improvement as compared to mono-dielectric HJLTFET—High-k HJLTFET and Low-k HJLTFET in terms of I_{ON}, I_{ON}/I_{OFF}, device efficiency, and SS. So, the combination of hetero-dielectric used in our proposed device tends to advance the analog performance of the device, which makes it an appropriate alternative for low power and fast switching applications.

Acknowledgements The authors would like to acknowledge the Microelectronics Research Lab, Department of Applied Physics, Delhi Technological University, to carry out this work.

References

1. Ghosh, B., Akram, M.W.: Junctionless tunnel field effect transistor. IEEE Electron Device Lett. **34**(5), 584–586 (2013)
2. Linienfeld, E.J.: Method and apparatus for controlling electric current. Patent first filed in Canada (1925)
3. Colinge, J.P., Lee, C.W., Afzalian, A., Akhavan, N.D., Yan, R., Ferain, I., … Kelleher, A.M.: Nanowire transistors without junctions. Nat. Nanotechnol. **5**(3), 225 (2010)
4. Ionescu, A.M., Riel, H.: Tunnel field-effect transistors as energy-efficient electronic switches. Nature, **479**(7373), 329–337 (2011)

5. Yu, E., Lee, W.J., Jung, J., Cho, S.: Ultrathin sige shell channel p-type finfet on bulk si for sub-10-nm technology nodes. IEEE Trans. Electron Devices **65**(4), 1290–1297 (2018)
6. Wong, J.C., Salahuddin, S.: Negative capacitance transistors. Proc. IEEE **107**(1), 49–62 (2018)
7. Zafar, S., D'Emic, C., Jagtiani, A., Kratschmer, E., Miao, X., Zhu, Y., ... Riel, H.: Silicon nanowire field effect transistor sensors with minimal sensor-to-sensor variations and enhanced sensing characteristics. Acs Nano. **12**(7), 6577–6587 (2018)
8. Madan, J., Gupta, R.S., Chaujar, R.: Performance investigation of heterogeneous gate dielectric-gate metal engineered–gate all around-tunnel FET for RF applications. Microsyst. Technol. **23**(9), 4081–4090 (2017)
9. Choi, W.Y., Park, B.G., Lee, J.D., Liu, T.J.K.: Tunneling field-effect transistors (TFETs) with subthreshold swing (SS) less than 60 mV/dec. IEEE Electron Device Lett. **28**(8), 743–745 (2007)
10. Madan, J., Chaujar, R.: Gate drain-overlapped-asymmetric gate dielectric-GAA-TFET: a solution for suppressed ambipolarity and enhanced ON state behavior. Appl. Phys. A **122**(11), 973 (2016)
11. Kumar, S., Goel, E., Singh, K., Singh, B., Singh, P.K., Baral, K., Jit, S.: 2-D analytical modeling of the electrical characteristics of dual-material double-gate TFETs with a SiO 2/HfO 2 stacked gate-oxide structure. IEEE Trans. Electron Devices **64**(3), 960–968 (2017)
12. Seabaugh, A.C., Zhang, Q.: Low-voltage tunnel transistors for beyond CMOS logic. Proc. IEEE **98**(12), 2095–2110 (2010)
13. Sharma, S., Chaujar, R.: Performance enhancement in a novel amalgamation of arsenide/antimonide tunneling interface with charge plasma junctionless-TFET. AEU-Int. J. Electron. Commun. **133**, 153669 (2021)
14. Convertino, C., Zota, C.B., Schmid, H., Ionescu, A.M., Moselund, K.E.: III–V heterostructure tunnel field-effect transistor. J. Phys. Condens. Matter **30**(26), 264005 (2018)
15. Vasen, T., Ramvall, P., Afzalian, A., Doornbos, G., Holland, M., Thelander, C., ... Passlack, M.: Vertical gate-all-around nanowire GaSb-InAs core-shell n-type tunnel FETs. Sci. Rep. **9**(1), 1–9 (2019)
16. Sharma, S., Chaujar, R.: Band gap and gate metal engineering of novel hetero-material InAs/GaAs-based JLTFET for improved wireless applications. J. Mater. Sci.: Mat. Electron. **32**(3), 3155–3166 (2021)
17. Choi, W.Y., Lee, W.: Hetero-gate-dielectric tunneling field-effect transistors. IEEE Trans. Electron Devices **57**(9), 2317–2319 (2010)
18. Hänsch, W., Vogelsang, T., Kircher, R., Orlowski, M.: Carrier transport near the Si/SiO_2 interface of a MOSFET. Solid-State Electron. **32**(10), 839–849 (1989)
19. ATLAS User's Guide, SILVACO Int., Santa Clara, CA, USA, Version 5.26.1.R (2018)
20. Zhao, Y., Wu, C., Huang, Q., Chen, C., Zhu, J., Guo, L., ..., Huang, R.: A novel tunnel FET design through adaptive bandgap engineering with constant sub-threshold slope over 5 decades of current and high I_{ON}/I_{OFF} ratio. IEEE Electron Device Lett. **38**(5), 540–543 (2017)

Superior Performance of Gate Workfunction and Gate Dielectric Engineered Trapezoidal FinFET in the Presence of Trap Charges

Priyanshi Goyal and Harsupreet Kaur

1 Introduction

In the last few decades, the microelectronic industry has grown enormously due to factors such as device miniaturization, improvements in fabrication techniques, and constant advancements in the area of novel materials. The shrinking of transistors not only improves packaging density but also increases the speed of operation. Nevertheless, device scaling also leads to deterioration in device performance due to undesirable effects such as threshold voltage roll-off, Drain Induced Barrier Lowering (DIBL), mobility degradation, velocity saturation, etc. Furthermore, in short channel devices, charges near the drain end get energized at very high drain voltages and few of these also get trapped in oxide layer. These hot carriers also degrade device performance. To overcome these effects, various multigate devices such as Double Gate (DG) MOSFET [1, 2], Surrounding gate MOSFET [3, 4], and FinFET [5] have been proposed. In such devices, the presence of multiple gates improves short channel immunity and also ensures high current driving ability. Apart from this, high-k materials [6] have also been introduced to reduce tunneling and to improve gate controllability. In view of the above, in the present work, an analytical model has been developed for the proposed device, Dual Material Gate—High-k Trapezoidal FinFET (DMG-HK Trapezoidal FinFET) that incorporates the effect of gate work-function and gate dielectric engineering. The impact of interface traps has also been accounted for in the model. It is anticipated that the presence of DMG design and high-k dielectric layer in the gate stack will help in overcoming various detrimental effects such as short channel effects and hot carrier effects. These devices also offer reduced power consumption compared to planar transistors and exhibit more drive current per unit area than planar devices. Furthermore, in order to critically assess the effectiveness of the proposed device, a comparison of the proposed device, Dual Material Gate—High-k Trapezoidal FinFET (DMG-HK Trapezoidal FinFET) has

P. Goyal · H. Kaur (✉)
University of Delhi South Campus, New Delhi 110021, India

A. Biswas et al. (eds.), *Microelectronics, Circuits and Systems*, Lecture Notes in Electrical Engineering 755, https://doi.org/10.1007/978-981-16-1570-2_2

been drawn with Conventional Trapezoidal FinFET (C-Trapezoidal FinFET) device as well as Dual Material Gate Trapezoidal FinFET (DMG—Trapezoidal FinFET) device. The device characteristics have been obtained by solving Poisson's equation and by employing appropriate boundary conditions. Using the model, various characteristics such as potential, electric field, threshold voltage, and drain induced barrier lowering (DIBL) have been obtained and these characteristics have been compared for all devices under consideration. The paper is structured as follows. In Sect. 2, model formulation has been presented. This is followed by results and discussion in Sect. 3 and finally conclusion is presented in Sect. 4.

2 Model Formulation

Figure 1 shows the schematic view of Dual Material Gate—High-k Trapezoidal FinFET (DMG-HK Trapezoidal FinFET). L_{g1} and L_{g2} denote the lengths of channel under gate (M1) with higher workfunction and gate (M2) with lower workfunction, respectively. The channel concentration is denoted by N_c. The fin height, top fin width, and bottom fin width are denoted as H_{fin}, W_{ft}, and W_{fb}, respectively.

In order to obtain device characteristics, the quasi 3D scaling equation is solved and is given as [7]

$$\frac{d^2\psi_j(z)}{dz^2} - \frac{1}{\lambda_1^2}\left(\psi_j(z) - \psi_j\right) = 0 \tag{1}$$

where $j = 1, 2$ corresponds to $(0 \le Z \le L_{g1})$ region 1 and $(L_{g1} \le Z \le L_g)$ region 2, respectively.

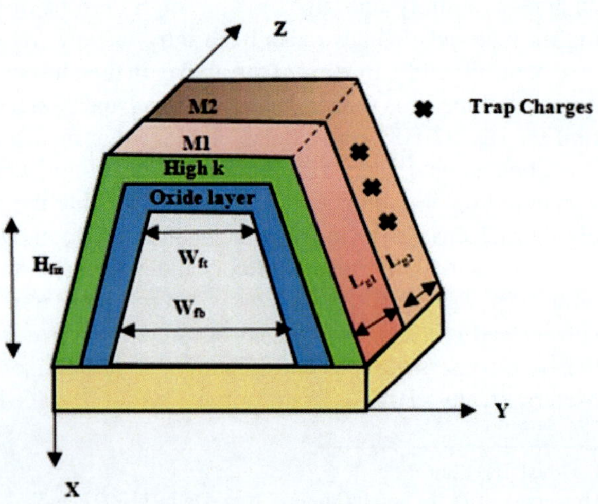

Fig. 1 Cross-sectional view of DMG-HK Trapezoidal FinFET

λ_l is scaling length of DMG-HK Trapezoidal FinFET, $\psi_j(Z)$ is the bottom central potential, ψ_j is the long channel bottom central potential and is given as

$$\psi_j = V_{gs} - V_{fb(j)} - \frac{qN_c}{\varepsilon_{si}}\lambda_l^2 \tag{2}$$

The flat band voltage at the source side is given as

$$V_{fb1} = \Phi_{M1} - \Phi_{Si} \tag{3}$$

Similarly, the flat band voltage at the drain side is given as

$$V_{fb2} = \Phi_{M2} - \Phi_{Si} - \frac{qN_{ft}}{C_{ox}} \tag{4}$$

where Φ_{M1}, Φ_{M2}, and Φ_{Si} are the workfunctions of metal 1 (near the source), metal 2 (near the drain end), and silicon, respectively. N_{ft} denotes trap charges and C_{ox} is oxide capacitance.

Now, in order to consider the effect of high-k layer in the gate stack, the effective oxide thickness is given as t_{oxeff} [8]

$$t_{oxeff} = t_l + \left(\frac{\varepsilon_{ox}}{\varepsilon_k}\right)t_2$$

where ε_k and ε_{ox} denote dielectric permittivity values of high-k layer and oxide layer, respectively, t_l is oxide layer thickness and t_2 is thickness of high-k layer.

Considering the boundary conditions and solving Eq. (1), the solutions for bottom central potential in regions 1 and 2 have been obtained as follows:

$$\psi_1(Z) = X_1 e^{\frac{Z}{\lambda_l}} + X_2 e^{\frac{-Z}{\lambda_l}} + \psi_1 \tag{5}$$

$$\psi_2(Z) = Y_1 e^{\frac{(Z-L_{g1})}{\lambda_l}} + Y_2 e^{\frac{-(Z-L_{g1})}{\lambda_l}} + \psi_2 \tag{6}$$

where

$$X_1 = x_1 V_{gs} + y_1$$

$$X_2 = x_2 V_{gs} + y_2$$

$$Y_1 = X_1 \exp\left(\frac{L_{g1}}{\lambda_1}\right) - \frac{(\psi_1 - \psi_2)}{2}$$

$$Y_2 = X_2 \exp\left(\frac{-L_{g1}}{\lambda_1}\right) - \frac{(\psi_2 - \psi_1)}{2}$$

$$x_1 = \frac{\left(\exp\left(\frac{-L_g}{\lambda_1}\right) - 1\right)}{2\sinh\left(\frac{L_g}{\lambda_1}\right)}$$

$$y_1 = V_{bi} + P_1 - y_2$$

$$x_2 = \frac{\left(1 - \exp\left(\frac{L_g}{\lambda_1}\right)\right)}{2\sinh\left(\frac{L_g}{\lambda_1}\right)}$$

$$y_2 = \frac{-V_{ds} + T_1 - (V_{bi} + P_2) - T_2}{2\sinh\left(\frac{L_g}{\lambda_1}\right)}$$

V_{gs}, V_{bi}, and V_{ds} denote gate to source voltage, built-in voltage, and drain voltage, respectively.

$$T_1 = (V_{bi} + P_1)\exp\left(\frac{L_g}{\lambda_1}\right)$$

$$T_2 = B_{21}\cosh\left(\frac{-L_{g2}}{\lambda_1}\right)$$

$$P_1 = V_{fb1} + \frac{qN_c}{\varepsilon_{si}}\lambda_1^2$$

$$P_2 = V_{fb2} + \frac{qN_c}{\varepsilon_{si}}\lambda_1^2$$

$$B_{21} = (\psi_2 - \psi_1)$$

Threshold voltage can be obtained by equating the minimum bottom potential to twice of bulk potential as follows:

$$\psi_{min}(Z) = 2\Phi_b \tag{7}$$

where minimum central potential, $\psi_{min}(Z)$ is given as

$$\psi_1(Z) = X_1 e^{\frac{Z}{\lambda_1}} + X_2 e^{\frac{-Z}{\lambda_1}} + \psi_1 \tag{8}$$

From above Eq. (2), (5), and (7)

$$2\varphi_b - V_{gs} + P_1 = X_1 e^{\frac{Z}{\lambda_1}} + X_2 e^{\frac{-Z}{\lambda_1}} \tag{9}$$

Substituting X_1 and X_2 in Eq. (9) and replacing V_{gs} with V_{th}, the following expression for V_{th} is obtained:

$$V_{th} = \frac{M - \sqrt{M^2 - QN}}{Q}$$

where

$$M = 2(x_1 y_2 + x_2 y_1) + (P_1 + 2\Phi_b)$$

$$Q = 1 - 4x_1 x_2$$

$$N = P_1^2 - 4y_1 y_2$$

Furthermore, subthreshold swing can be obtained using the following expression

$$SS = 2.3V_t \frac{1}{\frac{d\%_{min}}{dV_{gs}}} \tag{10}$$

where V_t is thermal voltage and $\psi_{min}(Z)$ is as given by Eq. (8).

3 Results and Discussion

In this section, using the derived model, parameters such as surface potential, electric field, threshold voltage, and subthreshold swing have been obtained for different values of trap charges. The parameters chosen in the model have been obtained by first calibrating the results for fresh DMG Trapezoidal FinFET with the results reported in [7] in the presence of strain. The parameters so obtained were then incorporated in the analytical model to obtain the results for the proposed device (DMG-HK Trapezoidal FinFET) in the presence of trap charges. A comparison has also been drawn out between Conventional—Trapezoidal FinFET (Device 1), DMG Trapezoidal FinFET (Device 2) and DMG-HK Trapezoidal FinFET (Device 3) for different trap charges. The impact of variation in gate length ratio of high and low workfunction regions as well as variation in length of damaged zone on device characteristics has also been analyzed. The physical thickness of oxide for Conventional Trapezoidal FinFET and DMG Trapezoidal FinFET is kept same as gate stack thickness of DMG-HK Trapezoidal FinFET.

Table 1 List of parameters

Parameters	Value
Top fin width (W_{ft})	15 nm
Bottom fin width (W_{fb})	20 nm
Fin height (H_{fin})	20 nm
Metal 1 workfunction (Φ_{M1})	4.9 eV
Metal 2 workfunction (Φ_{M2})	4.4 eV
Oxide thickness (t_1)	1 nm
High-k layer thickness (t_2)	2 nm
Channel concentration (N_C)	10^{16} cm^{-3}
Device length (L_g)	30 nm

In Table 1, the various parameters used in the present work are listed and these values have been used throughout the analysis unless otherwise specified.

Figure 2a–e shows the calibration of the results obtained using the derived model with the results reported in [7] and a good agreement can be observed.

Figure 3a–c shows the change in surface potential with position along channel for all devices in the absence of trap charges (fresh devices), and for all devices in the presence of positive and negative trap charges, respectively. The positive trap charges shift the potential upwards resulting in decrease in potential barrier. However,

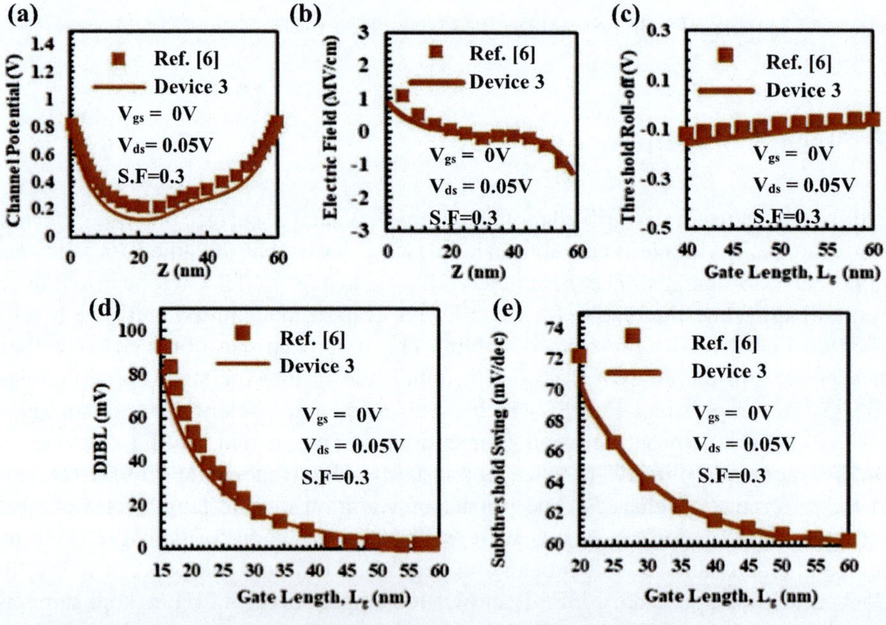

Fig. 2 **a–e** Calibration of analytical results with [7]. **a** Channel potential. **b** Electric field. **c** Threshold roll-off. **d** DIBL. **e** Subthreshold

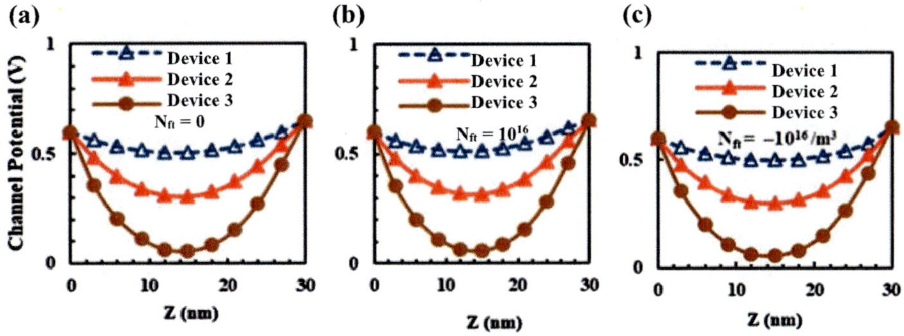

Fig. 3 **a–c** Channel potential along channel for all devices **a** for $N_{ft} = 0$, **b** $N_{ft} = 10^{16}/m^3$, **c** $N_{ft} = -10^{16}/m^3$, $L_d = 5$ nm, $V_{ds} = 0.05$ V, $V_{gs} = 0$ V

negative trap charges lead to the increase in barrier as compared to fresh device. On comparing Fig. 3a–c, it can be seen that the reduction in barrier is minimal for device 3 and the maximum reduction in barrier can be seen for device 1. Thus, it can be deduced that the presence of DMG and high-k layer leads to improvement in gate controllability which is responsible for increased barrier in DMG-HK Trapezoidal FinFET and this makes this device more immune to short channel effects.

Figure 4a–c shows the electric field along channel length for all devices in the absence of trap charges (fresh devices), and for all devices in the presence of positive and negative trap charges, respectively. The presence of positive trap charges leads to increase in electric field at drain side. On the other hand, the negative trap charges result in reduction in electric field near drain end. Moreover, the fact that device 3 exhibits lowest electric field values at drain end points to the fact that device 3 exhibits better resilience to hot carrier effects. Furthermore, device 3 exhibits a higher electric field value at the interface of the two regions which implies that device 3

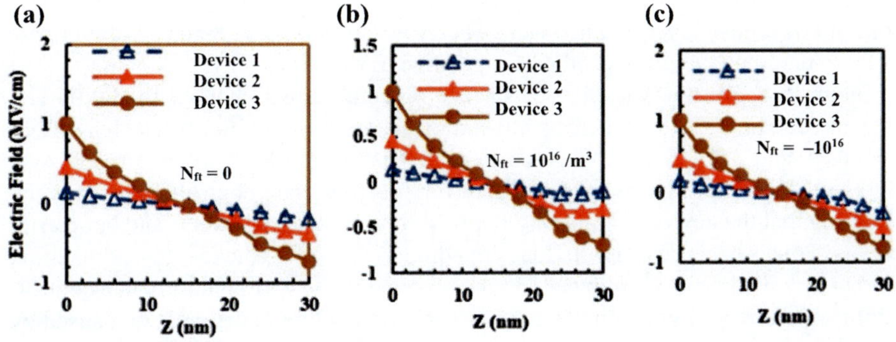

Fig. 4 **a–c** Electric field along channel for all devices **a** for $N_{ft} = 0$, **b** $N_{ft} = 10^{16}/m^3$, **c** $N_{ft} = -10^{16}/m^3$, $L_d = 5$ nm, $V_{ds} = 0.05$ V, $V_{gs} = 0$ V

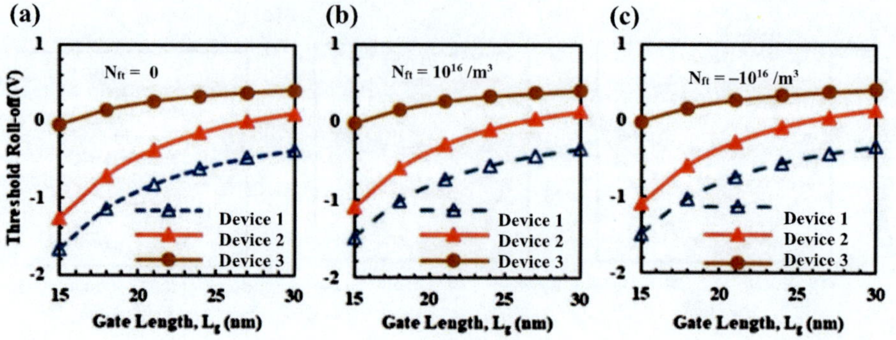

Fig. 5 **a–c** Threshold roll-off with gate length for all devices **a** for $N_{ft} = 0$, **b** $N_{ft} = 10^{16}/m^3$, **c** $N_{ft} = -10^{16}/m^3$, $L_d = 5$ nm, $V_{ds} = 0.05$ V, $V_{gs} = 0$ V

offers improved transport efficiency apart from improved capacitive coupling due to the presence of high-k layer in the gate stack.

Figure 5a–c shows the change in threshold voltage with gate length for all devices in the absence of trap charges (fresh devices), and for all devices in the presence of positive and negative trap charges, respectively. It can be observed that the presence of positive trap charges leads to degradation in threshold voltage due to decreased barrier whereas, negative trap charges lead to less degradation as compared to fresh device due to an increased barrier as was apparent from Fig. 3a–c. It is also clear from figure that DMG-HK Trapezoidal FinFET exhibits significant improvement in device performance as compared to conventional device, and device with only DMG design as the shift in threshold voltage is minimum for the proposed device which also implies that DMG-HK Trapezoidal FinFET exhibits less leakage current as compared to other two devices.

Since positive trap charges lead to degradation in device performance, it is necessary to analyze whether device performance can be improved by optimizing the lengths of regions with high and low workfunctions and the value of dielectric constant of high-k layer in gate stack. Therefore, in Figs. 6, 7, 8 and 9), the impact of these parameters will be studied for positive trap charges only.

Figure 6 shows the variation of surface potential along channel for DMG-HK Trapezoidal FinFET for fixed positive trap charges for different gate lengths of regions 1 and 2. It can be observed that increase in the length of damaged zone (L_d) leads to a decrease in barrier which implies poor gate controllability, thereby implying that the device becomes more prone to short channel effects and hot carrier effects. This is because as L_{g1} reduces in comparison to L_{g2}, the length of the region with higher workfunction reduces which consequently leads to reduced gate controllability, thereby implying that device becomes more prone to degradation caused by trap charges.

Figure 7 shows the variation of threshold voltage with gate length for DMG-HK Trapezoidal FinFET for fixed positive trap charges for different gate length ratios. It can be noticed that a smaller L_{g1} value leads to more degradation in threshold

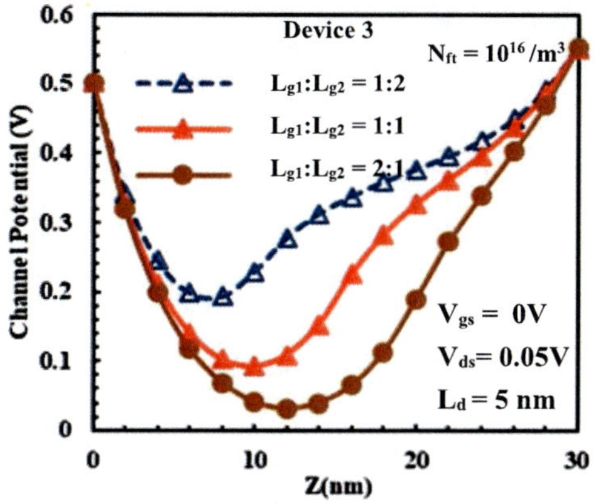

Fig. 6 Channel potential along channel for DMG-HK Trapezoidal FinFET for different gate length ratios

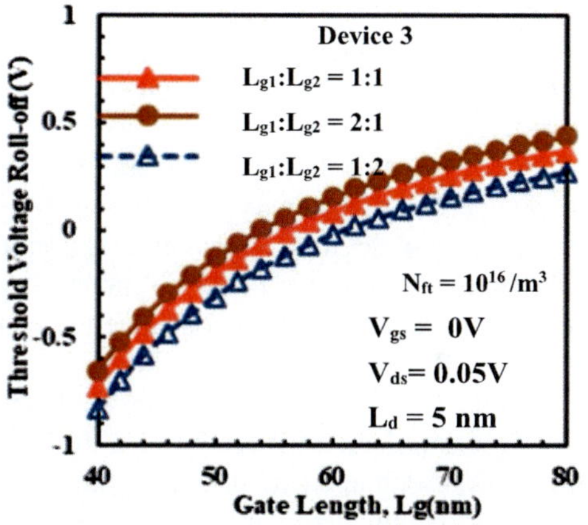

Fig. 7 Threshold voltage with gate length for DMG-HK Trapezoidal FinFET for gate length ratios

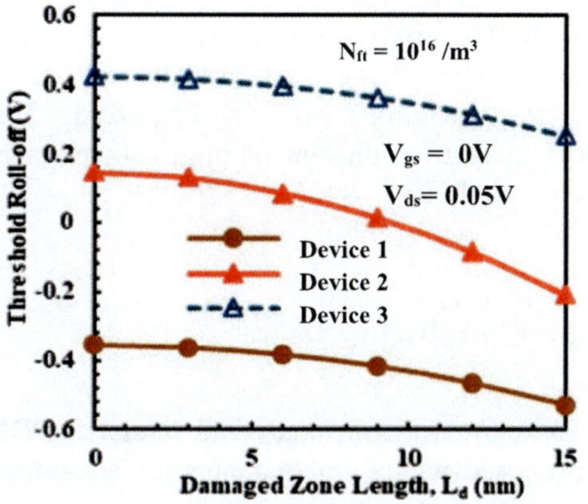

Fig. 8 Threshold roll-off with damaged zone zone length for all devices

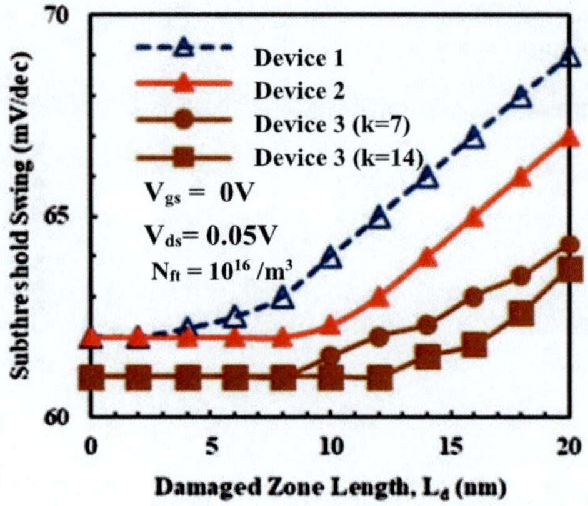

Fig. 9 Subthreshold swing with damaged length for all devices

voltage. This is because the decrease in length of region with higher metal workfunction leads to reduced gate control, thereby resulting in degraded threshold voltage characteristics.

Figure 8 demonstrates the threshold voltage roll-off with damaged zone length for all devices. It can be analyzed that trap charges shift the threshold voltage and cause degradation in device performance. It is evident that the degradation in voltage roll-off becomes more effective with increased damaged zone length. Furthermore, the impact of trap charges and consequently change in damaged zone length is minimal on device 3.

In Fig. 9, the variation of subthreshold swing with damaged zone length has been shown for all devices. It can be seen that subthreshold swing degrades with increase in length of damaged zone. Furthermore, the impact of trap charges and increase in length of damaged zone has least impact in case of DMG-HK Trapezoidal FinFET as compared to Conv-Trapezoidal FinFET and DMG Trapezoidal FinFET because DMG-HK offers improved capacitive coupling and gate controllability which leads to less degradation due to trap charges.

Furthermore, DMG-HK Trapezoidal FinFET offers improved performance as the dielectric constant of high-k layer increases because of enhancement in gate controllability.

4 Conclusion

To summarize, an analytical model for DMG-HK Trapezoidal FinFET has been proposed which amalgamates the advantages of DMG design and high-k layer in gate stack. The model takes into account the effect of trap charges and in order to analyze the efficacy of the proposed device in suppressing the degradation caused by trap charges, various characteristics such as channel potential, electric field, threshold voltage roll-off, DIBL, and subthreshold swing have been obtained. An exhaustive comparison of these characteristics has been performed for all devices to analyze the efficacy of DMG-HK Trapezoidal FinFET in overcoming the degradation caused due to trap charges. It has been demonstrated that even in the presence of trap charges, DMG-HK Trapezoidal FinFET offers improved gate controllability as well as better immunity to short channel effects and hot carrier effects thereby implying better resilience of DMG-HK Trapezoidal FinFET towards trap charges.

References

1. Fasarakis, N., Karatsori, T.A., Tsormpatzoglou, A.T., Tassis, D.H., Papathanasiou, K., Bucher, M., Ghibaudo, G., Dimitriadis, C.A.: Compact modeling of nanoscale Trapezoidal Fin FETs. IEEE Trans. Electron Devices (2014)
2. Mangesh, S., Chopra, P.K., Saini, K.K.: Analysis of extended pile gate Trapezoidal Bulk FinFET. IETE J. Res. (2019). https://doi.org/10.1080/03772063.2019.1579678
3. Aouaj, A., Bouziane, A., Nouacr, A.: Analytical Vth and S models for (DMG-GC stack) surrounding-gate MOSFET. Int. J. Electron. **99**(1) (2012)
4. Auth, C.P., Plummer, J.D.: Scaling theory for cylindrical fully depleted, surrounding gate MOSFETs including effective conducting path effect (ECPE). Microelectron. Eng. **77**, 175–183 (2005). https://doi.org/10.1016/j.mee.2004.10.005
5. Wu, X., Chan, P.C.H.: Impacts of nonrectangular Fin cross section on the electrical characteristics of FinFET. IEEE Trans. Electron Devices **52**(1) (2005)
6. Narendar, A., Mishra, R.A.: Analytical modeling and simulation of multigate FinFET devices and the impact of high-k dielectrics on short channel effects (SCEs). Superlatt. Microstruct. **85**(2015), 357–369 (2015)
7. Banerjee, P., Sarkar, S.K.: Comprehensive analysis of subthreshold short channel behavior of a dual-material gate strained Trapezoidal FinFET. Superlatt. Microstruct. (2018). https://doi.org/10.1016/j.spmi.2018.02.034
8. Haddara, H., Cristoloveanu, S.: Two-dimensional modeling of locally damaged short-channel MOSFET's operating in the linear region. IEEE Trans. Electron Devices **34**(2) (1987)

High-K Biomolecule Sensor Based on L-Shaped Tunnel FET

Prarthana Chakraborti, Abhijit Biswas, and Abhijit Mallik

1 Introduction

Over the last few decades FET-based sensors have been extensively investigated due to their numerous advantages including simplicity in fabrication, scalability, compatibility with standard CMOS technology, low cost, high sensitivity, and also label-free detection process [1]. Any good sensing devices must possess some important characteristics such as high sensitivity, low detection limit, large dynamic range, rapid and real-time detection, low cost and reduced size [2–4]. In order to meet these requirements, various kinds of Metal Oxide Field Effect Transistors (MOSFETs) have emerged as sensing devices in recent years [2, 5]. Of them asymmetric double gate transistors, impact ionization FET [6], underlap channel embedded FET, dielectric modulated (DM)-FET, DMTFET endowed with novel architectural modifications like gate underlap, split gate, and many more have been used as sensors to detect biomolecules using their physical properties such as dielectric constant and charge density [1, 4]. In particular, DMFETs have become popular for working in an only dry environment for the detection of different types of neutral and charged biomolecules. Some of the selected biomolecules are responsible for pulmonary congestion, dizziness, nausea, drowsiness, reduced level of consciousness, and others are used in both biochemical and pharmaceutical applications. Biosensing in a wet environment

P. Chakraborti (✉) · A. Biswas
Department of Radio Physics and Electronics, University of Calcutta, APC Road, Kolkata 700009, West Bengal, India
e-mail: prarthana3133@gmail.com

A. Biswas
e-mail: abiswas5@rediffmail.com

A. Mallik
Department of Electronic Science, University of Calcutta, APC Road, Kolkata 700009, West Bengal, India
e-mail: abhijit_mallik1965@yahoo.co.in

A. Biswas et al. (eds.), *Microelectronics, Circuits and Systems*, Lecture Notes in Electrical Engineering 755, https://doi.org/10.1007/978-981-16-1570-2_3

by using L-shaped Tunnel FETs is new as well as practical due to the fact that most of the biomolecules are collected from blood or urine sample for diagnostic purposes. But, the ionic nature of solvent, for example, water affects the sensitivity of biomolecules because sometimes the ions of biomolecules may be screened by ions of solvent. In recent years silicon-based immunological FET (ImmunoFET) comprising an ISFET in which the gate dielectric is functionalized with ligands to detect the targeted analyte is being used in CMOS industry. However, it faces some challenges while operating in liquid environment. So, some modifications like floating gate immunoFET, DG immunoFET, DM immunoFET are introduced to mitigate these challenges [7]. DMDG FETs are more immune to short channel effects, however, with comparatively less sensitivity but exhibiting appreciable sensitivity roll-off [8]. More recently, UDCM-SON-FET with no unwanted interfacial layer is used to sense mainly the charged biomolecules whereas for neutral biomolecules it reports the sensitivity of 1.17 V [9].

Among the different kinds of FET structures, steep slope transistors such as tunnel FETs [7–12] are very interesting as sensing devices since they exhibit the subthreshold slope below the lower limit of 60 mV/decade at room temperature. Although there have been numerous findings on the use of TFETs as sensors, impacts of various kinds of TFET architectures like U-shape, V-shape, L-shape [13], the concepts of carrier in-line tunneling with gate field and also the biomolecules having high dielectric constant have not yet been addressed for biosensing application.

This paper presents a comprehensive study on sensing of different biomolecules with high dielectric constant by using L-shaped tunnel FETs in wet environment. In this paper, we consider neutral biomolecules which have higher dielectric constant values ranging 12.3–46.7 exceeding the SiO_2 value, i.e., 3.9 and such biomolecules are referred to as high-k materials.

The work relies on carrier in-line tunneling concepts where the line tunneling produces more I_D and steeper SS compared to lateral tunnel FET. All results are obtained using well-calibrated SILVACO ATLAS device simulator [14]. Also, the total gate electric field being strong enough can ignore the screening effect of charges in a solvent and detect biomolecules from the variations in the device characteristic that occurs due to the impact of biomolecules on the tunneling mechanism in the source-channel junction. Moreover, our results on sensitivity are compared with the earlier data reported in the literature.

This paper is organized as follows: Sect. 2 describes the device structure along with the simulation framework. Section 3 deals with model calibration results. Section 4 describes the simulation results and explanations of the results followed by drawing concluding remarks in Sect. 5.

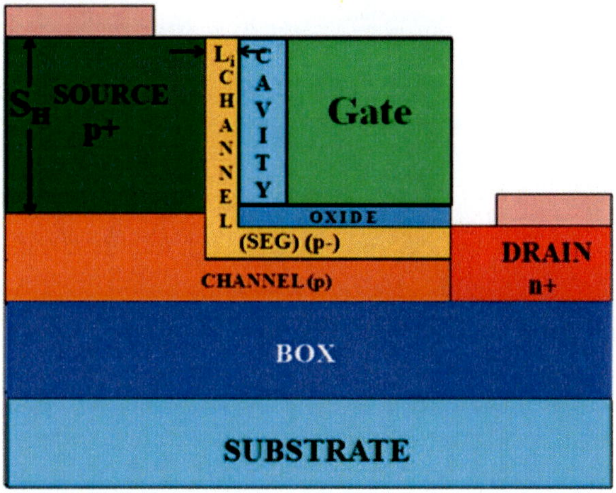

Fig. 1 Schematic of device structure with doping concentration of Source (p+) = $1 \times 10^{20}/cm^3$, Channel (SEG) (p−) = $1 \times 10^{17}/cm^3$, Channel (p) = $1 \times 10^{15}/cm^3$ and Drain (n+) = $1 \times 10^{19}/cm^3$, respectively

2 Device Structure and Simulation Framework

2.1 Device Description and Structure

The proposed L-shaped n-Tunnel FET for sensing various biomolecules is shown in Fig. 1. The detailed fabrication steps of such a device may be found in [13]. 200 nm thick buried oxide layer is formed on the top of the Si substrate on which a 60 nm thick p-type Si channel having a concentration of $1 \times 10^{15}/cm^3$ is placed. p-type source of 200 nm height and n-type drain of 60 nm height are doped with acceptor concentration of $1 \times 10^{20}/cm^3$, donor concentration of $1 \times 10^{19}/cm^3$, respectively. The p-type L-shaped selective epitaxial growth (SEG) region is working as the channel [13] having doping concentration of $1 \times 10^{17}/cm^3$. HfO_2 with physical thickness of 6 nm is used as the gate oxide. Some portion of HfO_2 is etched for making a cavity in which target biomolecules having a particular dielectric constant can attach to the receptors at the oxide-semiconductor interface. TiN having work function of 4.1 eV is employed as the gate material and Gold is used to form source and drain contacts.

2.2 Simulation Setup

SILVACO ATLAS device simulator [14] is used to simulate TFETs in order to obtain current-voltage characteristics. For FET-based biosensors the primary and important work is to find the proper area for sensing the biomolecules and to model those biomolecules, so that the device can capture the sensing effects through modification of its drain current. This work considers wet condition analysis of various biomolecules like HEWL, FABP, LLAMA antibody, dimethyl sulfoxide, pyridine,

Table 1 Dielectric constant of different biomolecules using water as the solvent

Sl. No.	Names of biomolecules	Dielectric constant
1	HEWL (Hen Egg White Lysozyme)	25.7
2	FABP (Rat Fatty Acid Binding Protein)	40
3	LLAMA (antibody)	17.2
4	Dimethyl sulfoxide	46.7
5	Pyridine	12.3
6	Methanol	33.64

and methanol and their dielectric constants are listed in Table 1. Some amounts of biomolecules are captured in the cavity formed by etching the gate oxide and attached to the receptors adhered to the interface of oxide and semiconductor of the device. Bandgap narrowing model, Shockley-Read-Hall recombination model, Fermi-Dirac statistics, and Trap assisted Tunneling models are used in the simulation. In order to capture the tunneling current nonlocal band-to-band tunneling (BTBT) model combined with CVT Lombardi mobility model is employed.

3 Model Calibration

Our simulated characteristic curve is compared with the experimental transfer characteristic of n-channel LTFET reported in [13]. The experimental TFET consists of 70 nm high p-type Si source with doping concentration of $1 \times 10^{20}/\text{cm}^3$, p-type Si mesa shaped Selective Epitaxial Growth (SEG) region serving as the channel with doping concentration of $1 \times 10^{17}/\text{cm}^3$ and n-type Si drain region with doping concentration of $1 \times 10^{18}/\text{cm}^3$. 6 nm thick HfO_2 is used as the gate oxide in which L-shaped structure was sculpted. The gate comprises a metal with work function of 4.52 eV. A good agreement is observed between the experimental and simulated characteristics (Fig. 2), which validates our simulation framework.

4 Results and Discussions

4.1 Effects of Source Height on Drain Current

The height of the source influences the drain current versus gate voltage characteristics as shown in Fig. 3. As the source height rises the effective tunneling area of the device increases thereby enhancing the tunneling rate of electrons from the valence band of the source to the conduction band of the channel. Consequently drain

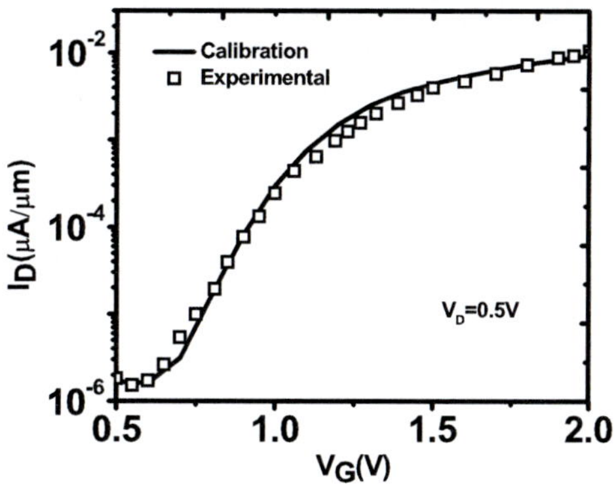

Fig. 2 Model calibration against experimental data in [13]

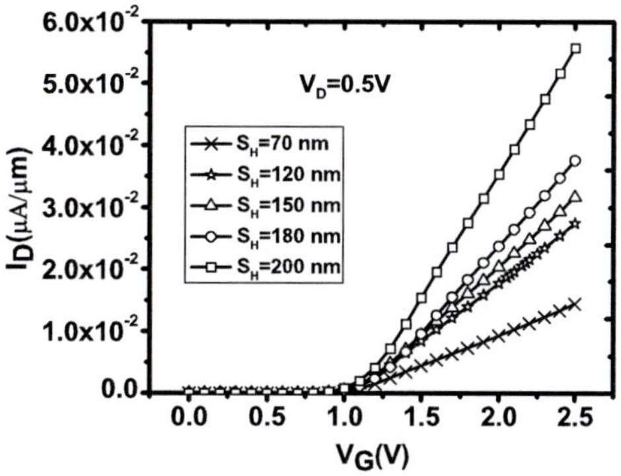

Fig. 3 I_D–V_G characteristics with the variation of source height

current shows improvement with increasing source height for a given gate voltage as demonstrated in Fig. 3. Notably, as the source height increases from 70 to 200 nm, the drain current shows an increment of 5 times at $V_G = 2.5$ V.

4.2 *Effects of Thickness of SEG Region (L_i) on Drain Current*

Figure 4 shows the variation of drain current with gate voltage for $L_i = 9$ and 10 nm. As L_i of the device is reduced from 10 to 9 nm, the tunneling window becomes narrower facilitating more electrons to tunnel through the narrower tunneling window as compared to the number of carriers for $L_i = 10$ nm. The enhanced drain current for $L_i = 9$ nm relative to the current for $L_i = 10$ nm is clearly shown in Fig. 4 for various values of V_G. As it is known that at least 10 nm space is required to fit biomolecules perfectly in the cavity, Fig. 5 shows the transfer characteristics curve at $L_i = 9$ nm with $T_{OX} = 10$ nm. Hence, the sensing device is designed with $S_H = 200$ nm, $L_i =$

Fig. 4 Variation of I_D with V_G for $L_i = 9$ nm and $L_i = 10$ nm with $T_{OX} = 6$ nm

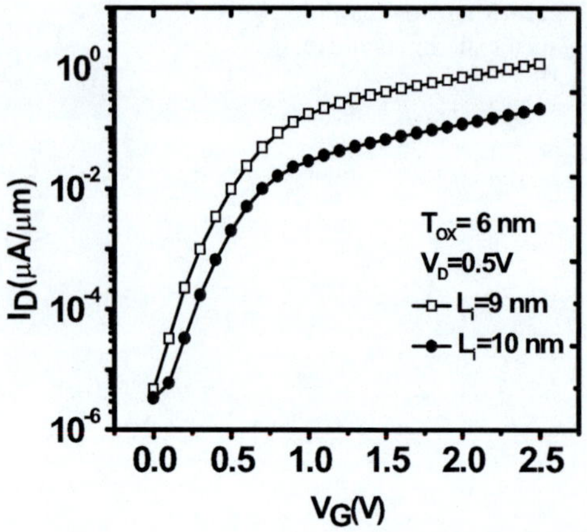

Fig. 5 Variation of I_D with V_G for $T_{OX} = 6$ nm and 10 nm at $L_i = 9$ nm

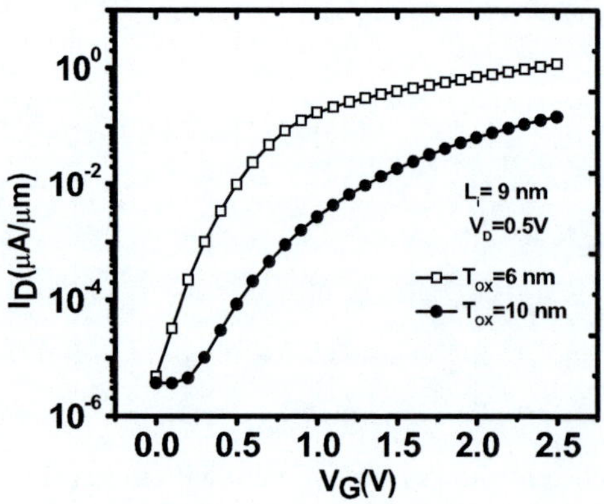

9 nm, and $T_{OX} = 10$ nm filling the cavity with water having dielectric constant of almost $80\varepsilon_0$ [5].

4.3 Sensing of Biomolecules

Figure 6a, b shows the variation of drain current with respect to the gate voltage in the range from 0 to 2.5 V with and without biomolecules. Most importantly, it may be noticed that the threshold voltage of the device increases considerably as it contains biomolecules only. Figure 7a shows the variation of gate voltage with dielectric constant of various biomolecules mentioned in Table 1 at five constant I_D values. A particular species, characterized by a definite dielectric constant, can be detected by the steep change in V_G among a vast range of biomolecules. It is worthwhile noting that various biomolecules produce steeper change in V_G for I_D

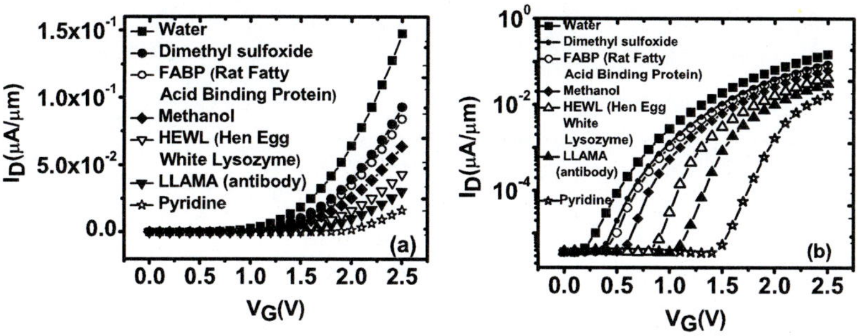

Fig. 6 I_D-V_G characteristic curve in **a** linear and **b** semi-log scale for different biomolecules having various permittivity values

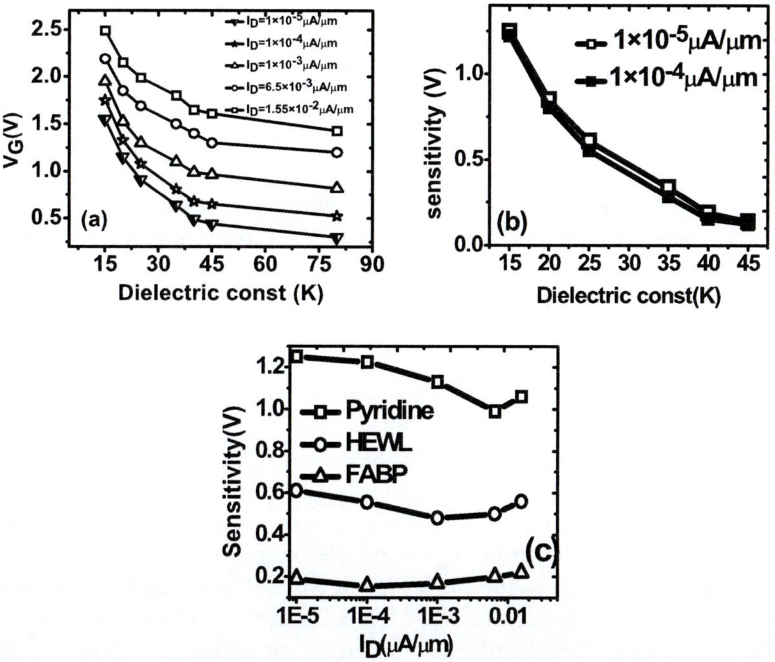

Fig. 7 **a** variation of gate voltage with dielectric constant of different biomolecules and water **b** variation of voltage sensitivity with dielectric constant of different biomolecules and water and **c** variation of voltage sensitivity with drain current

$= 1 \times 10^{-5}$and 1×10^{-4} μA/μm (Fig. 7a). The voltage sensitivity, defined by $|V_G(K = \text{water}) - V_G(K = \text{Biomolecules})|$, is evaluated and plotted against dielectric constant of various biomolecules for constant I_D of 1×10^{-5} and 1×10^{-4} μA/μm in Fig. 7b. Intriguingly, biomolecules having comparatively low dielectric constant can be detected with comparatively higher sensitivity of almost 1.3 V whereas the

Table 2 Performance comparison of biosensors

Sl. No	Name of device	Sensitivity
1	Nanoscale FET-based biosensor	22 mV [15]
2	SOI TFET	90 mV [16]
3	SOI MOSFET	300 mV [16]
4	DM DG JL MOSFET	400 mV [17]
5	DM DG MOSFET	0.45–0.35 V [8]
6	UDCM SON FET	1.17 V [9]
7	DG NW FET	1.3 V [18]
8	This work	1.3 V

biomolecules having high dielectric constant show comparatively low sensitivity of almost 150 mV. As sensitivity of various biomolecules is affected by the drain current, the variation of sensitivity is plotted against I_D for three different biomolecules FABP, HEWL, pyridine in Fig. 7c.

The sensitivity of our proposed TFET for various biomolecules is compared with the sensitivity values reported for biomolecules reported elsewhere in Table 2. It is noticed that this work has achieved almost the same value or higher value of sensitivity as compared to earlier findings. However, all the previous works were performed in dry environment which is usually used to sense biomolecules of low dielectric constant.

5 Conclusion

We have presented a comprehensive study on sensing of various biomolecules having high dielectric constant in wet environment using L-shape TFET sensors. The performance of the sensor is influenced by the variation of source height and width of the Selective Epitaxial Growth (SEG) region and the optimized sensor is designed with suitable choice of source height and width of SEG. The evaluated sensitivity for pyridine is in the range from 1 to 1.3 V, for HEWL from 474 to 600 mV, and for FABP ranging from 155 to 220 mV as drain current varies from 10^{-2} to 10^{-5} μA/μm. These sensitivity values are either better or comparable to findings reported earlier.

Acknowledgements The first author acknowledges CSIR, HRDG, India for providing fellowship for SRF vide File No. 09/028(1028)/2018-EMR-I dtd. 16.04.2018.

References

1. Ajay, N.R., Saxena, M., Gupta, M.: Drain current model of a four-gate dielectric modulated MOSFET for application as a biosensor. IEEE Trans. Electron. Device **62**, 2636–2644 (2015)
2. Bergveld, P.: Thirty years of ISFET Technology What happened in the past 30 years and what may happen in the next 30 years. Sens. Actuators. B. Chem. **88**, 1–20 (2003)
3. Kim, Y.J., Ahn, H.J., Choi, J.S., Im, M., Kim, S., Duarte, P.J., Kim, H.C., Park, J.T., Lee, Y.S., Choi, K.Y.: An underlap channel-embedded field-effect transistor for biosensor application in watery and dry environment. IEEE Trans. Nanotech. **11**, 390–394 (2012)
4. Lee, W.K., Choi, J.S., Ahn, H.J., Moon, I.D., Park, T.J., Lee, S.Y., Choi, K.Y.: An underlap field-effect transistor for electrical detection of influenza, Appl. Phys. Lett. **96**,1–3 (2010)
5. Ajay, N.R., Saxena, M., Gupta, M.: Analytical model of pH sensing characteristics of junctionless silicon on insulator ISFET. IEEE Trans. Electron. Device **64**, 1742–1750 (2017)
6. Bandiziol, A., Palestri, P., Pittino, F., Esseni, D., Selmi, L.: A TCAD based methodology to model the site-binding charge at ISFET/electrolyte interfaces. IEEE Trans. Electron. Devices **62**, 3379–3386 (2015)
7. Bhattacharyya, M.I., Cohen, S., Shalabny, A., Bashouti, M., Akabayov, B., Shalev, G.: Specific and label-free immunosensing of protein-protein interactions with silicon-based immune FETs. Biosens. Bioelectron. **132**, 143–161 (2019)
8. Balakrishnan, B., Balamurugan, B.N.: 2D analytical modeling and simulation of dual material DG MOSFET for biosensing application, AEU Int. J. Electron. Commun. **99**, 193– 200 (2019)
9. Singh, N.K., Dutta, K.P.: Analytical modeling of a high–K underlap dielectric and charge–modulated silicon on nothing FET based biosensor. J. Comput. Electron. **19**, 1126–1135 (2020)
10. Ionescu, M.A., Riel, H.: Tunnel field effect transistors as energy efficient electronic switches. Nature **479**, 329–337 (2011)
11. Omura, Y., Mori, Y., Sato, S., Mallik, A.: Revisiting the role of trap-assisted-tunneling process on current-voltage characteristics in tunnel field-effect transistors. J. Appl. Phys. **123**, 1–6 (2018)
12. Mori, Y., Sato, S., Omura, Y., Mallik, A.: Proposal of the possible method to define the threshold voltage of lateral tunnel field-effect transistors. Technol. Rep. Kansai Univ. (Osaka, Japan) **59**, 75–81 (2017)
13. Kim, W.S., Kim, H.J., Liu, K.J.T., Choi, Y.W., Park, G.B.: Demonstration of L-shaped tunnel field effect transistors. IEEE Trans. Electron. Devices **63**, 1774–1777 (2016)
14. ATLAS User's Manual: Silvaco. Santa Clara, CA, USA (2013)
15. Shoorideh, K., Chui, O.C.: On the origin of enhanced sensitivity in nanoscale FET-based biosensors. *PNAS* **111**, 5111–5116 (2014)
16. Dubey, A., Narang, R., Saxena, M., Gupta, M.: Investigation of total ionizing dose effect on SOI tunnel FET. Superlattices Microstruct. **133**, 49–54 (2019)
17. Rahman, E., Shadman, A., Khosru, M.D.Q.: Effect of biomolecule position and fill in factor on sensitivity of a Dielectric Modulated Double Gate Junction less MOSFET biosensor. Sens. Bio-Sens. Res. **13**, 49–54 (2017)
18. Ahn, H.J., Choi, J.S., Han, W.J., Park, J.T., Lee, Y.S., Choi, K.Y.: Double-Gate nanowire field effect transistor for a biosensor. Nano Lett. **10**, 2934–2938 (2010)

A Novel Trench FinFET as a Biosensor for Early Detection of Alzheimer's Disease

M. Aditya and Rishu Chaujar

1 Introduction

The ability to determine health status, onset and development of illness, and track treatment outcome through a non-invasive approach is the key goal to be accomplished in the promotion and provision and study of healthcare. To achieve this goal, there are three prerequisites: unique biomarkers suggesting a stable or diseased state; a non-invasive method for detecting and tracking the biomarkers [1].

In recent years, demand for simple and portable devices in the field of medical diagnostics has grown, that also have quick response times, are user-friendly, cost-effective, and are suitable for mass production [1]. Biosensor technologies offer the ability to meet those requirements through an interdisciplinary synthesis of nanotechnology, chemistry, and medical science approaches [2, 3].

Detection of biomolecules is one of the main research areas of concern these days; understandably since the detection of biomolecules will help identify diseases such as Alzheimer's disease [3]. The prompt diagnosis of Alzheimer's helps with disease management and slowing its progression. Alzheimer's disease is caused by the production and deposition of the β-amyloid peptide/protein (βA) on the sulci on the surface of the brain. The solubility of βA, and the quantity of βA, are closely related to disease progression.

βA monomers can agglomerate into many types of assemblies, like oligomers, protofibrils, and fibrils. Fibrils (3–7 nm) [4] are largest and insoluble, and they can further assemble into plaques, while amyloid-oligomers are soluble and spread throughout the cerebrospinal fluid and brain [4].

M. Aditya (✉) · R. Chaujar
Department of Applied Physics, Delhi Technological University, Delhi, India
e-mail: adityam98@gmail.com

R. Chaujar
e-mail: chaujar.rishu@dtu.ac.in

A. Biswas et al. (eds.), *Microelectronics, Circuits and Systems*, Lecture Notes in Electrical Engineering 755, https://doi.org/10.1007/978-981-16-1570-2_4

High sensitivity, reliability, and ease of use are the fundamental for various biological and biomedical applications. Biosensors based on field transistor (FET) have attracted attention in recent years due to easy-scalability, high-sensitivity, fast and direct electrical response, minimal power usage, and cheaper cost compared to alternative methods [5].

The variation in the devices electrical properties with the introduction of βA inside the device cavity causes the device to work as a biosensor. In our trench FinFETs, when βA is introduced in the trench of the device, the dielectric constant of βA can change the on and off current of the device [6, 7]. We propose this device to detect the presence of βA oligomers in the cerebrospinal fluid of the patient.

In this work we present a novel Trench FinFET for the detection of βA peptide for the early detection and diagnosis of Alzheimer's disease.

2 Device Structure

The structure of the device is shown in Figs. 1 and 2. The structural constraints of the device are given in Table 1. The device was modeled using SILVACO DEVEDIT 3D Structure and Mesh Editor. The device is a FinFET whose fin is trenched in the middle to accommodate biomolecules inside. Figure 3 shows the proposed fabrication steps to fabricate the device.

The trench on the fin facilitates the βA to lodge themselves inside thus allowing them to be detected. The cavity is selected according to the diameter of the strands

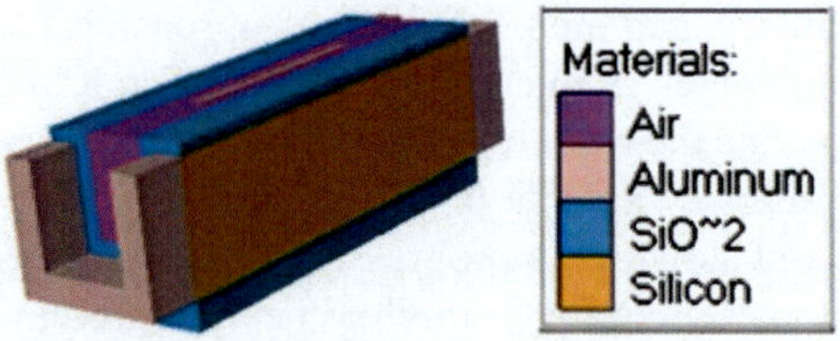

Fig. 1. 3D view of the simulated Trench FinFET

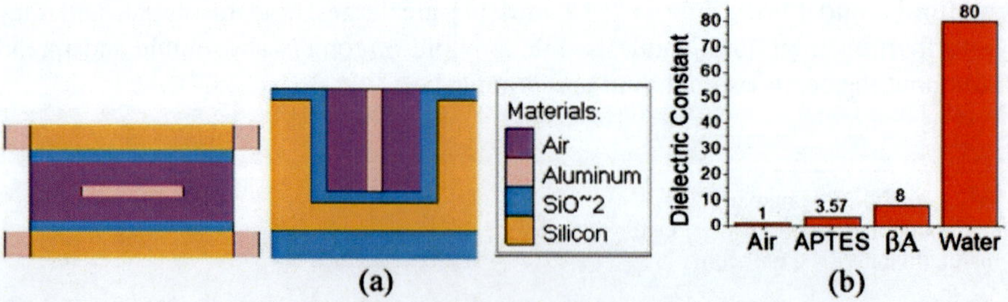

Fig. 2 **a** Lateral and transverse section of the device. **b** Dielectric constants of the various molecules

Table 1 Device structure parameters

Parameter	Value
Channel length (L_g)	40 nm
Fin height (H_f)	23 nm
Fin width (W_f)	26 nm
Thickness oxide (t_{ox})	2 nm
Trench width	5 nm
Trench depth	18 nm
Source and drain range	20 nm
Source and drain doping	10^{21} cm^{-3}
Gate work function	4.26 eV

(3–7 nm) to be 5 nm [4] and the length is selected as the average length of βA fibril strands is 60–220 nm [8] while the monomers hexamers and oligomers are smaller. Thus this device can be used to detect βA inside the trench of the device. The βA is bound to the trench using APTES as a binding molecule.

The device has a fin 80 nm in length that is covered on the surface by 2 nm thick Silicon Dioxide with a 40 nm Aluminum gate. The measurements of the fin were 23 nm × 26 nm. The measurements of the trench are 80 nm × 18 nm × 5 nm. The fin doping was a uniform n-type doping of 10^{15} cm^{-3} while the source and drain n-type doping were of 10^{21} cm^{-3}.

Device modeling was done using SILVACO ATLAS Device Simulator using the Fermi–Dirac Statistics, Concentration Dependent Arora Mobility Model, and Concentration Dependent SHR Recombination Model [9, 10].

3 Methodology

The dielectric constant variation of the analyte inside the device trench and the subsequent variation in the electrical properties of the device is the primary analysis methodology that has been applied in this work.

This method is used to imitate the habitation of βA in the interior of the trench allowing the device to act as a sensor. The dielectric constant of βA is used. The dielectric constant of water is taken to be 80. Most proteins have an internal dielectric constant of 2–8 [5]. Proteins with high molecular masses (100–200 kDa) such as βA have an internal dielectric constant close to 8 and high polarizability due to the presence of large numbers of peptide linkages within them [4]. A value of 2–8 for the dielectric constant is used to model various proteins static internal dielectric constant and that approach is applied here to simulate βA. The size of βA fibrils can range from 3 to 7 nm in diameter [4, 11], while oligomers can range from small hexamers to large spheroid structures (12–200 *monomers*). Thus a trench width of 5 nm is simulated here taking into consideration that the average width of the

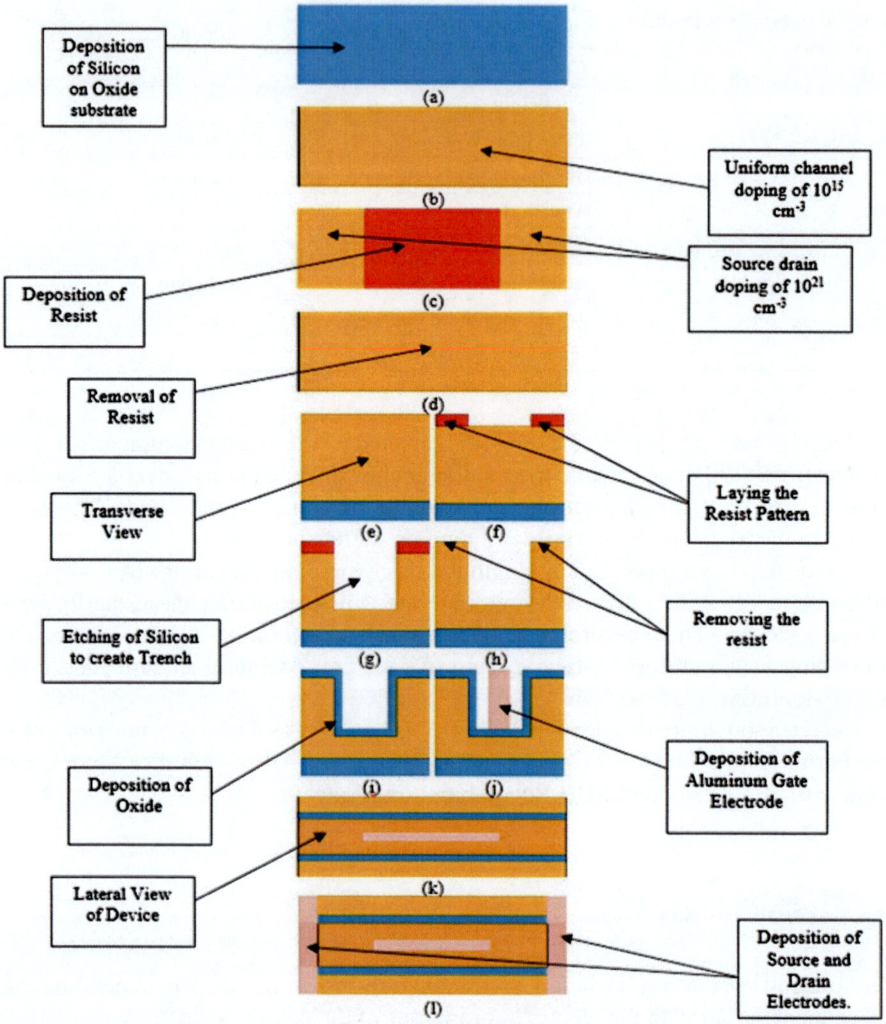

Fig. 3 **a** Deposition of silicon on oxide **b** uniform channel doping **c** deposition of resist and then source drain doping **d** removal of resist **e** transverse view **f** laying the resist pattern **g** etching of silicon to create trench **h** removing the resist **i** deposition of oxide **j** deposition of aluminum gate electrode **k** lateral view of device **l** deposition of source and drain electrodes

fibrils is 5.5 nm which is more than the size of the βA oligomers [11]. APTES ((3-Aminopropyl) triethoxysilane) is used as a binding molecule to bind the βA peptide and oligomers inside the trench [12]; APTES has a dielectric constant of 3.57 which is low compared to βA peptide due to lower polarizability of the APTES molecule, we have also simulated the response of the device in presence of only APTES to give a clearer picture of the devices working. The change in the electrical characteristics

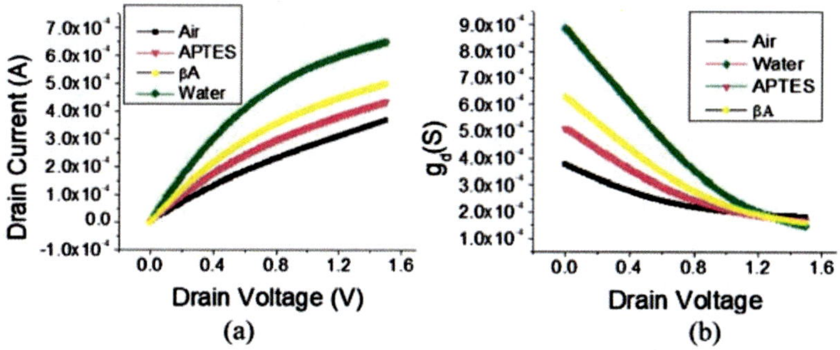

Fig. 4 **a** I_{ds}*versus* V_{gs} characteristics of the device at various molecules at $V_{gs} = 1.5\ V$. **b** g_d at different molecules at $V_{gs} = 1.5\ V$

of the device in presence of βA peptide is then used to detect the presence of βA peptide in the analyte [13-17].

4 Results and Discussions

4.1 *Output Characteristics*

Figure 4 presents the I_{ds}*versus* V_{ds} characteristics simulated for βA in the trench at $V_{gs} = 1.5\ V$. The I_{on}/I_{off} and S were derived from these characteristics. Equation (1) shows the expression used to calculate the sensitivity.

$$S = \frac{I_{on(k>1)}}{I_{on(k=1)}} \tag{1}$$

Subsequently, I_{ds} can be seen to increase with the increase in the relative permittivity (***k***), caused by the rise in the effective $\overrightarrow{E}$ in the interior of the fin causing superior channel generation. The sensitivity of the device also shows an increasing trend with relative permittivity.

4.2 *Transfer Characteristics*

Figure 5 presents the I_{ds}*versus* V_{gs} characteristics for βA at $V_{ds} = 0.5\ V$. The threshold shift (ΔV_{th}) from air is also presented. The increase in I_{ds} is ascribed to the increased permeation of the $\overrightarrow{E}$ applied on the fin due to the presence of βA modeled by the relative permittivity of the trench. The rise in the relative permittivity of the trench

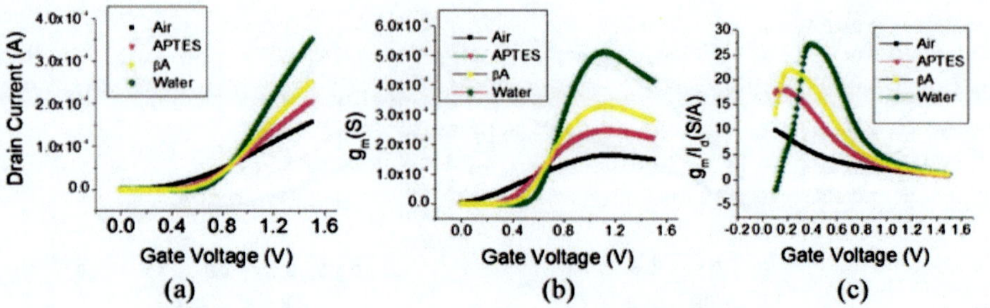

Fig. 5 **a** I_{ds} *versus* V_{gs} characteristics of the device for various molecules at $V_{ds} = 0.5\ V$. **b** g_m for various molecules at $V_{ds} = 0.5\ V$. **c** g_m/i_d of the device at $V_d = 0.5\ V$

causes a decrease in the effective width thus increasing C_{gs} resulting in superior channel generation.

This causes a movement in the V_{th} of the device. The g_m/i_d *versus* V_g of the device is also shown and it is observed that it increases with the presence of βA peptides inside the trench, this is attributed to the increase in the g_m of the device in the presence of βA peptides.

4.3 Electric Field Distribution

Figure 6 presents the $\vec{E}$ Distribution where it can be seen that the $\vec{E}$ in the trench rises with the rise in relative permittivity keeping V_{gs} constant caused by the fall in the effective oxide width when βA is inside the trench. The rise in the effective permittivity of the trench in the presence of βA peptides causes the effective oxide width to decrease thus increasing C_{gs}. The rise in the $\vec{E}$ in the fin due to better penetration of the V_{gs} through the oxide layer thus resulting in superior channel generation in presence of βA peptides.

4.4 Electron Current Distribution

Figure 7 presents the electron current density. The current density follows a rising trend with the relative permittivity of the biomolecule inside the trench. This rise can be ascribed to the rise in the $\vec{E}$ inside the trench due to the presence of βA peptides. It is also observed that the channel width on the fin surface increase thus increasing the current at higher relative permittivity ($I_{on(k>1)}$), this is also due to the presence of βA peptides and the better channel forming capability of the device due to an increase in $\vec{E}$ in the interior of the fin. This causes an increase in sensitivity in the presence of βA peptides.

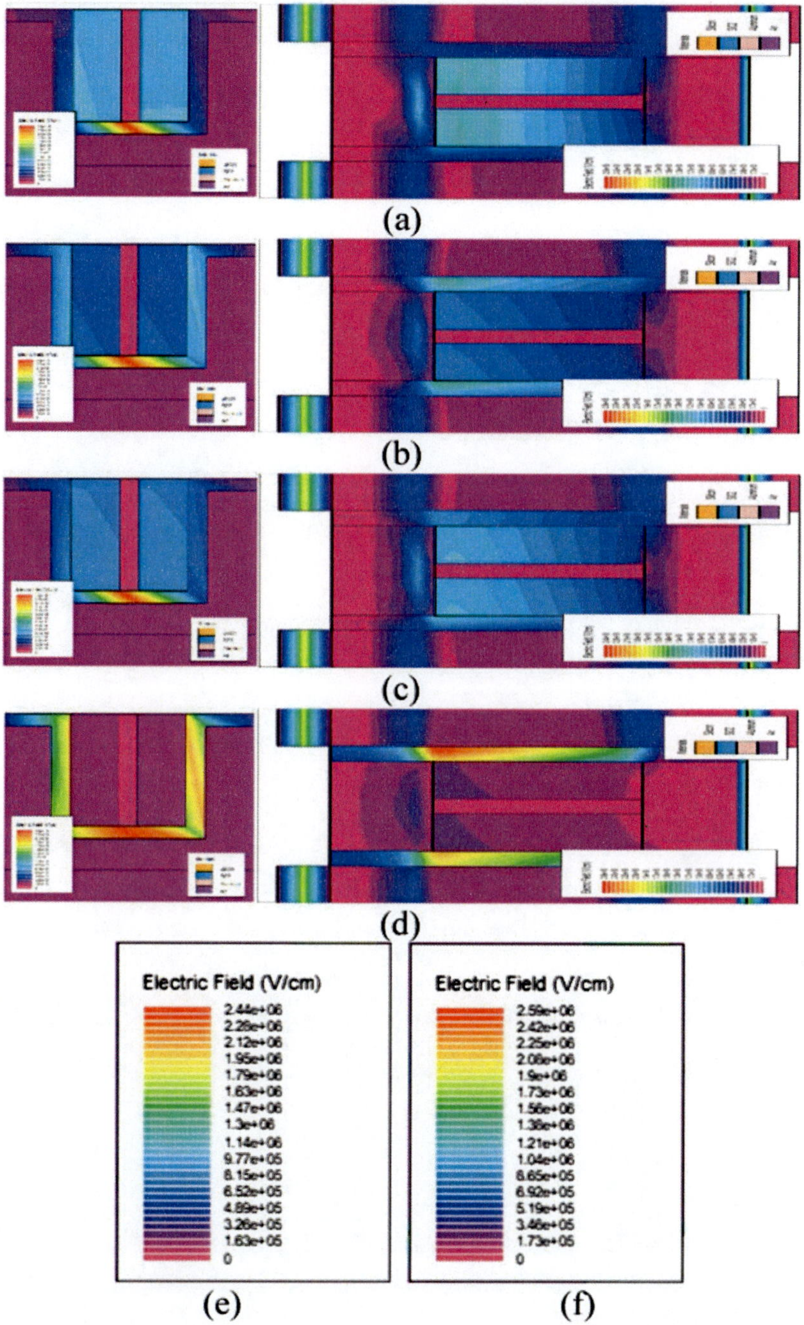

Fig. 6 Electric field **a** Air **b** βA **c** APTES **d** Water **e** Transverse E scale **f** longitudinal E scale

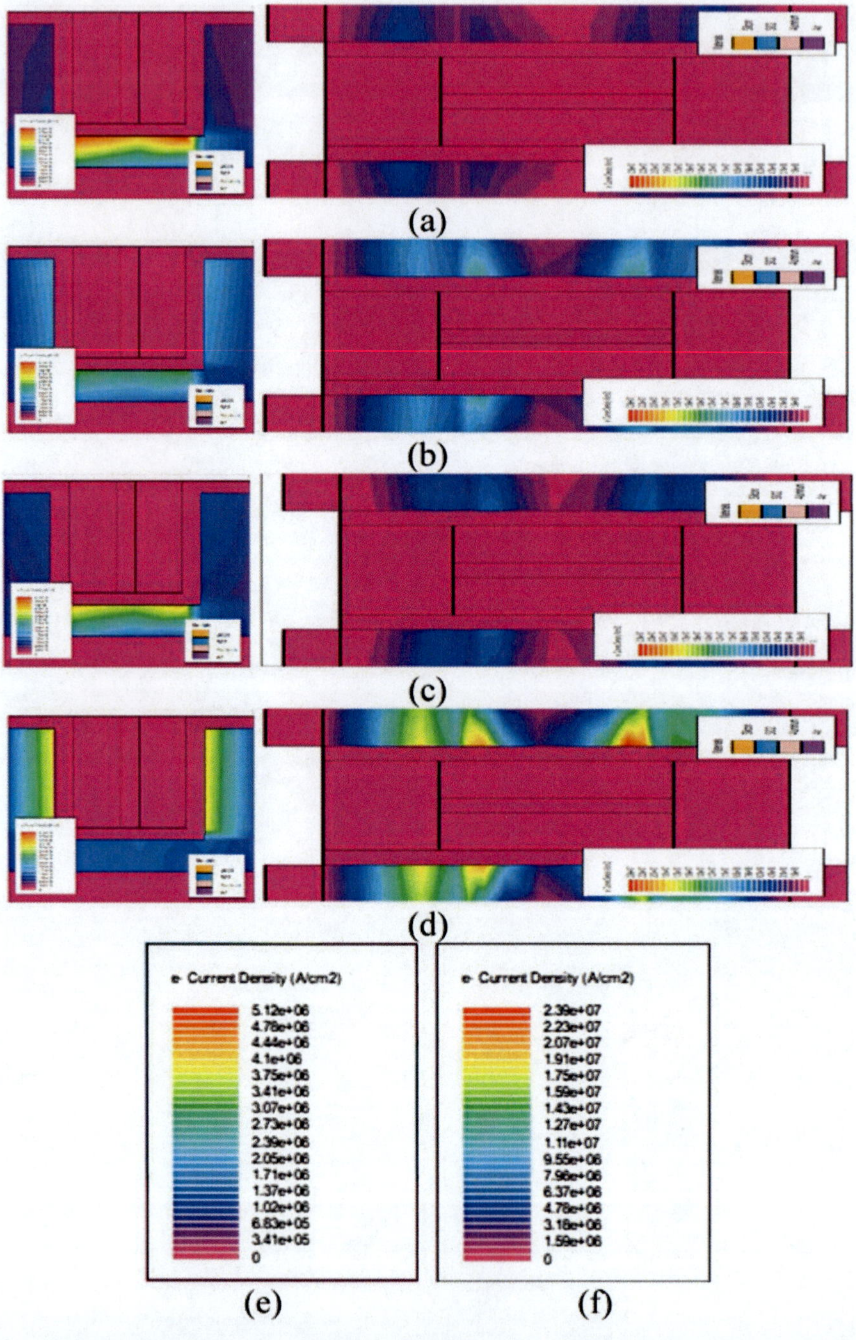

Fig. 7 Electron current density **a** air **b** βA **c** APTES **d** water **e** transverse e^- scale **f** longitudinal e^- scale

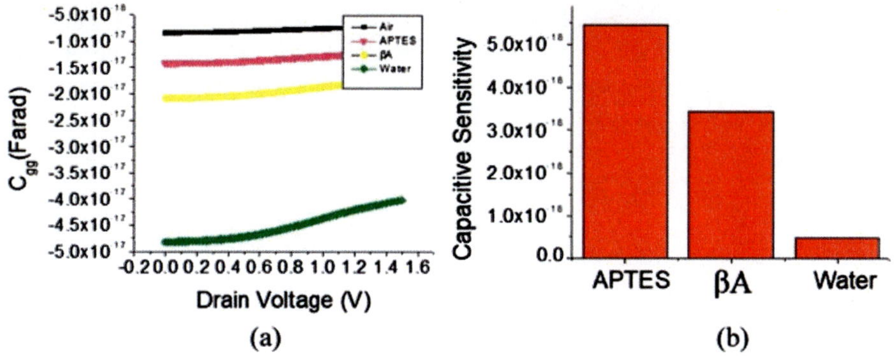

Fig. 8 **a** C_{gg} at $V_{ds} = 0.5$ V. **b** Capacitive sensitivity

Table 2 Sensitivity parameters

Medium	V_{th} (V)	ΔV_{th} (V)	Sensitivity	I_{on}/I_{off}
Air	0.392	0	1	26.74
APTES	0.556	0.164	1.18	2635.54
βA	0.616	0.224	1.25	14,666.66
Water	0.679	0.287	1.78	15,966.85

4.5 *Sensitivity and Threshold Voltage Shift*

The sensitivity defined in Eq. (1) and ΔV_{th} are shown in Table 2. The ΔV_{th} shift due to the presence of the molecule in the trench is calculated with respect to air. It is seen that the Shift increases with an rise in the relative permittivity of the molecule in the trench, this can be ascribed to the rise in the $\vec{E}$ in the cavity which causes a rapid rise in the current in the device thus increasing the slope of the curve. This increase in the slope consequently increases the threshold voltage thus increasing the shift.

$$SR = \frac{I_{on}}{I_{off}} \tag{2}$$

The switching ratio defined in Eq. (2) increases due to the rapid increase in the $I_{on} of$ the device with the rise in the relative permittivity of the molecules.

4.6 *Capacitance and Capacitive Sensitivity*

It can be observed from Fig. 8 that the C_{gg} of the device increases in magnitude with the presence of βA in the trench. This can be explained by the rise in the $\vec{E}$ in the interior of the fin due to the effective oxide thickness decreasing as the dielectric constant of the trench increases. Thus allowing higher penetration of the electric field inside the oxide layer thus increasing the capacitance of the device. Equation 3 shows the capacitive sensitivity of the device.

$$S_c = \frac{\Delta C_{gs}}{k} \tag{3}$$

5 Conclusion

Presented in this work is an original Trench FinFET for sensing of βA that can be used for early diagnosis of Alzheimer's disease. The analysis involves the variation in the response of the sensor and changes in the device parameters when the trench is filled with βA. APTES acts as a binding molecule and the dielectric constant in the trench changes. The ΔV_{th} follows a rising trend with relative permittivity and $\Delta V_{th\,max}$ of $0.616\ V$ for βA. The $I_{ds\,ON}$ of the device also rises from 3.64×10^{-4}A in air to 4.98×10^{-4}A for βA. The I_{on}/I_{off} of the device also rises from 26.74 in air to 1.466×10^{4} in βA. The $\vec{E}$ and e^{-} current density in the fin interior also increases with relative permittivity. The sensitivity of the device is 1.28 for βA. These changes in the electrical response of the device are identified to be primarily caused by the variations in the electrical parameters due to the presence of βA from the basis of its bio-sensing action.

Acknowledgements This work was supported by Delhi Technological University. The author is grateful for the support provided from the Microelectronics Lab for this research.

References

1. Comini, E., Baratto, C., Faglia, G., Ferroni, M., Ponzoni, A., Zappa, D., Sberveglieri, G.: Metal oxide nanowire chemical and biochemical sensors. J. Mater. Res. **28**(21), 2911–2931 (2013)
2. Lee, J., Jang, J., Choi, B., Yoon, J., Kim, J.-Y., Choi, Y.-K., Kim, D.M., Kim, D.H., Choi, S.-J.: A highly responsive silicon nanowire/amplifier MOSFET hybrid biosensor. Sci. Rep. **5** 12286 (2015)
3. Bergveld, P.: The development and application of FET-based biosensors. Biosensors **2**(1), 15–33 (1986)
4. Chen, G., Xu, T., Yan, Y., et al.: Amyloid beta: structure, biology and structure-based therapeutic development. Acta Pharmacol. Sin **38**, 1205–1235 (2017). https://doi.org/10.1038/aps.2017.28
5. Li, L., Li, C., Zhang, Z., Alexov, E.: On the 'dielectric "constant" of proteins: smooth dielectric function for macromolecular modelling and its implementation in DelPhi. J. Chem. Theory Comput. **9**, 2126–2136 (2013). https://doi.org/10.1021/ct400065j
6. Sang, S., Wang, Y., Feng, Q., Wei, Y., Ji, J., Zhang, W.: Progress of new label-free techniques for biosensors: a review. Crit. Rev. Biotechnol. **36**(3), 465–481 (2016)

7. Im, H., Huang, X.-J., Gu, B., Choi, Y.-K.: A dielectric-modulated field-effect transistor for biosensing. Nat. Nanotechnol. **2**(7), 430–434 (2007)
8. Murphy, M., LeVine, H.: Alzheimer's disease and the β-amyloid peptide. J. Alzheimer's Dis.: JAD. **19**, 311–323 (2010). https://doi.org/10.3233/JAD-20101221
9. Drive, P.H., Clara, S., Atlas User's Manual, Device Simulation Software, pp. 1– 1715 (2017).
10. Drive, P.H., Clara, S.: DEVEDIT User's Manual, Device Structure Editing Software, pp. 1–1715 (2017)
11. Dubnovitsky, A., Sandberg, A., Rahman, M., Benilova, I., Lendel, C., Härd, T.: Correction: amyloid-β protofibrils: size, morphology and synaptotoxicity of an engineered mimic. PLoS ONE **8**, e66101 (2013). https://doi.org/10.1371/journal.pone.0066101
12. H-N, Chan, X. Di, S-L. Ho, D. He, M. Wong, H-W. Li, : Highly sensitive quantification of Alzheimer's disease biomarkers by aptamer-assisted amplification. Theranostics. **9**, 2939–2949 (2019). https://doi.org/10.7150/thno.29232
13. Lau, H., Bae, T.-E., Jang, H.-J., Kwon, J.-Y., Cho, W.-J., J-OLim. : Saliva-based screening approach for Alzheimer's disease via the cell-oriented ion-sensitive field-effect transistor. Sensor Letters. **12**, 1096–1101 (2014). https://doi.org/10.1166/sl.2014.3200
14. Tripathi, M.M., Chaujar, R.: Sub-30nm In2O5Sn gate electrode recessed channel MOSFET: a biosensor for early stage diagnostics. Vacuum **164**, 46–52 (2019). https://doi.org/10.1016/j.vacuum.2019.02.054
15. Vu, X.T., Eschermann, J.F., Stockmann, R., GhoshMoulick, R., Offenhäusser, A., Ingebrandt, S.: Top down processed silicon nanowire transistor arrays for biosensing. Phys. Status Solidi **206**(3), 426–434 (2009)
16. Ahangari, Z.: Performance assessment of dualmaterial gate dielectric modulated nanowire junctionless MOSFET for ultrasensitive detection of biomolecules. RSC Adv. **6**(92), 89185–89191 (2016)
17. Chandan, B.V., Nigam, K., Sharma, D.: Junctionless based dielectric modulated electrically doped tunnel FET based biosensor for label-free detection. Micro Nano Lett. **13**(4), 452–456 (2018)

Interface Trap Charge Analysis of Junctionless Triple Metal Gate High-k Gate All Around Nanowire FET-Based Biotin Biosensor for Detection of Cardiovascular Diseases

Mekonnen Getnet Yirak and Rishu Chaujar

1 Introduction

Recently, the primary area of research interest is reduced semiconductor device size to satisfy the need for smaller integrated circuits (ICS) with the same functionality and increase the number of ICs in a single chip for better device performance [1]. As the size of technology is continually decreasing, the number of transistors in a single chip and transistors operation speeds are increases [2]. However, hugely scaling down device dimensions arises various problems the so-called short channel effects (SCEs), such as mobility degradation, threshold voltage roll-off, low on-current, high off-current, hot carrier, and impact ionization effect, parasitic resistance/capacitances, DIBL, and a substrate with defects, are worsening to achieve improved device performance with better reliability [3, 4]. Another problem of CMOS transistor is p–n junction related problem, which requires expensive fabrication techniques [5, 6] due to diffusion of impurities between p- or n-type drain/source region and n- or p-type body region pose great difficulty in the production of short channel/small scale devices. To continue Si CMOS device scaling, down to ultra-small device dimension, and to suppress device scaling-related problems, different approaches have been examined. For instance, a novel multi-gate structure such as double metal gat or surround gate MOS [7], replacing SiO_2 gate oxide by high-k dielectric material [8, 9], gate engineering, metal gate rather than polysilicon gate [10], and junctionless transistors having uniform doping profile from source to drain through the channel region [11] have been designed. Effective oxide thickness (EOT) can be scaled down to an ultra-small device dimension using a high-k gate dielectric material without increasing gate tunneling current [12].

M. G. Yirak (✉)
Physics Department, Debre Tabor University, Debre Tabor, Ethiopia

R. Chaujar
Applied Physics Department, Delhi Technological University, Delhi, India
e-mail: chaujar.rishu@dtu.ac.in

A. Biswas et al. (eds.), *Microelectronics, Circuits and Systems*, Lecture Notes in Electrical Engineering 755, https://doi.org/10.1007/978-981-16-1570-2_5

Similarly, replacing the polysilicon gate with a metal gate electrode (MG), depletion effect caused by the polysilicon gate is reduced [13]. Replacing polysilicon gate by metal gate electrode reduces parasitic capacitance and depletion effects caused by a polysilicon gate. But impact of interface charges trapped at buried oxide/channel interfaces due to hot carrier effects carried by the fabrication process, plasma etching, and high electric field effect along with ultra-small device dimension, arise defects/impurities or damage the gate oxide, are not well addressed. These will be serious issues on device performance [14–16], and Large traps at the interface of oxide/channel alter device electrostatics. Also, the impact of interface trap charges on the performance of junctionless surrounding gate all around nanowire-based biotin biosensor is not reported [17]. NWFET-based biosensor has become a primary focus for many researchers for the following reasons; these are easy scalability, cost-effective, label-free for detection, compatibility with bulk CMOS, high sensitivity, fruitful on-chip integration, less response time for detection and lower fabrication cost, and low power consumption [18–20]. The operation of biosensor technology lies in the fact of the successful translation of biological properties of the biomolecule to equivalent electronic properties. In case of NWFET-based biotin biosensor, the same action has been done by first finding the same electronic properties in dielectric constant and charge density and studding electrical parameters, such as threshold voltage, drain current, transconductance, surface potential, corresponding to the absence and presence of biomolecules in the NWFET-based biosensor. Critical challenges that enact the sensitivity of FET biosensors are DIBL [16], hot carrier effects, short channel effects, as a result of narrow the imperfect interface region on oxide/channel interface due to high electric field in case of short channel device [20].

In this article, n-type junctionless (JL) TM-high-k GAA—NWFET-based biotin biosensor has been proposed for the first time to study the impact of interface trap charges on device sensitivity by considering interface charges near or at the Si-SiO_2/channel, as shown in Fig. 1b.

2 Device Structure and Simulations

The device structure of n-type triple metal high-k gate all around junctionless NWFET-based biotin biosensor with interface trap charge used in this work has been illustrated in Fig. 1. Here, L_1 (6 nm), L_2 (8 nm), L_3 (6 nm) are the lengths of gate one, gate two, gate three, respectively, and L_4 is the length for both nanogap cavity and silicon dioxide (SiO_2), which are near to drain and source end respectively is (10 nm), L is channel length (20 nm). T_1, T_2, T_3, and T_4 are the thickness of metal gate, nanogap cavity, hafnium oxide, and interface (SiO_2) oxides, respectively, and 2R is the diameter of the channel. A 0.3 nm thickness of SiO_2 interface layer is considered between hafnium oxide and silicon film to create more compatible hafnium oxide with silicon film. The three gate materials are G_1, G_2, and G_3 having different work function denoted by $\Phi_{G1} = 4.86$, $\Phi_{G2} = 4.96$ and $\Phi_{G3} = 4.50$,

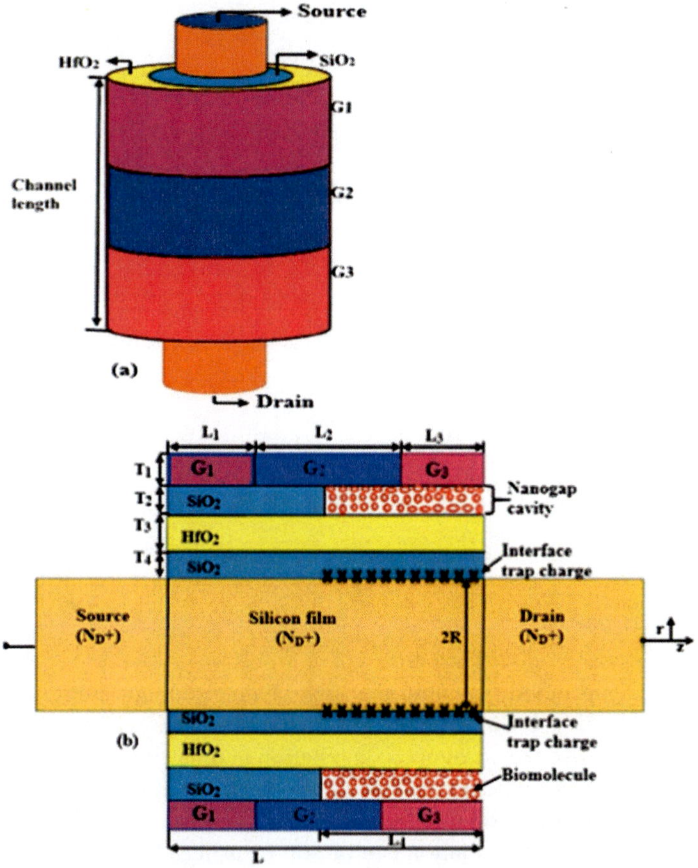

Fig. 1 **a** 3-D schematic structure and **b** 2-D cross-sectional view with cavity region for n-type triple metal gate high-k GAA-JL-NWFET

respectively. The work function (Φ_{G1}) near to the source is used to control electron saturation velocity, the work function (Φ_{G2}) between source end and drain end is used to control potential profile along the channel, which intern reduces SCEs and lower work function near to drain (Φ_{G3}) is used for screening effects [3–5]. HfO_2 is used to over com quantum mechanical tunneling [6]. The nanogap cavity region is used as a detecting site in which the target biomolecules are assumed to be uniformly immobilized in the nanogap cavity region. Biotin ($\epsilon = 2.63$), and air ($\in = 1$) cavity regions are considered by introducing their dielectric constant material [8]. Interface trap charge (ITCs) ($N_f = \pm 5 \times 10^{16}\,cm^{-2}$) and neutral charge ($N_f = 0$) are considered interchangeably for all simulations (see Table 1).

In this work, electrical properties of triple metal gate high-k GAA-junctionless NWFET device structure have been characterized using the "atlas 3-D" device simulator tool. Concentration-dependent mobility (CONMOB) [9] model was used

Table 1 Proposed device structural parameters

Parameters	TG-GAA-JL-NWFET		
Channel length (nm)	20		
Gate oxide thickness (nm)	$HfO_2 = 1.5$ and $SiO_2 = 0.3$		
Oxide thickness near to the source (nm)	$SiO_2 = 1$		
Oxide length near to the source (nm)	$SiO_2 = 10$		
Nanogap cavity length (nm)	10		
Source/Drain thickness (nm)	10		
Source/Drain length (nm)	10		
Nanogap cavity thickness (nm)	1		
The diameter of silicon (nm)	10		
Interface trap charges (ITCs)	$\pm 5 \times 10^{12}$ cm^{-2}		
Source/Drain and channel doping (N_D+)	10^{19} cm^{-3}		
Oxide dielectric constant	$SiO_2 = 3.9$ and $HfO_2 = 25.0$		
Gate Work functions (eV)	$\Phi_{G1} = 4.86$	$\Phi_{G2} = 4.96$	$\Phi_{G3} = 4.50$

to simulate bandgap mobility within the high channel doping along the bandgap narrowing (BGN) model. Shockley–Read–Hall model along with the Boltzmann transport model was used to account recombination of minority carriers [9]. Carrier–carrier scattering mobility model (CCSMOB) was used at higher carrier concentration. For parallel and perpendicular field-dependent mobility, CVT model has been used [10]. Carrier transport equation can be solved using Gummel's and Newton's numerical methods. But models of quantum mechanical effects have not been invoked in this simulation because the radius of the silicon film is not less than 4 nm [11]. Interface charge density ($N_f = \pm 5 \times 10^{12}$ cm^{-2}) has been introduced as localized charge at the Si–SiO_2 interface near to drain end.

3 Results and Discussion

In this study, output characteristics of triple metal gate high-k gate all around junctionless nanowire FET-based biotin biosensor has been examined by incorporating interface trap charges at Si–SiO_2/channel interface.

3.1 Effect of Different ITCs on Drain Current

Figure 3 demonstrates the simulated results of I_D-V_G transfer characteristics on log scale at $V_D = 0.05$ V with different interface trap charges, including air and biotin biomolecule for a triple metal gate. It is clear that higher leakage current for positive interface trap charges (ITCs) is observed compared to negative ITCs; this is because negative ITC reduces short channel effects and hot carrier effects while positive ITCs reduces charge carriers within the channel raises depletion layer to get thicker across the reverse junction that permits the flow of charge carriers in off-state. For biotin biomolecule, lower leakage is examined compared to air; the reason is that biotin biomolecule acts as high-k dielectric material and improves gate electrostatic control [13]. For instance, I_{OFF} for biotin biomolecule at ($N_f = \pm 5 \times 10^{12}\ \text{cm}^{-2}$) are 3.12×10^{-12} A and 4.07×10^{-16} A for positive and negative interface trap charge, respectively, at $V_{GS} = 0$ and $V_{DS} = 0.05$ V. JL-NWFET-based biotin biosensor is the device that uses biological molecules (biotin) to detect the presence of chemicals. Biotin biomolecule in FET device is an interface condition in the boundary as the link between the oxide of the FET and the analyte (an aqueous solution) which contains the bio-sample for purification and detection of various biomolecules and it is a water-soluble vitamin that functions as a prosthetic group in carboxylation reactions. In addition to its detection, biotin has multiple roles in gene regulation [14, 15]. Figure 2 demonstrates/validates the calibration simulation results of our proposed device.

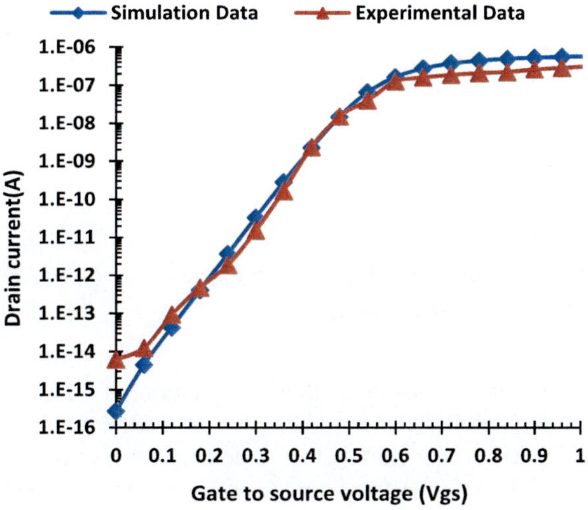

Fig. 2 Simulation and experimental calibration results [12] for n-type triple metal gate high-k GAA-JL-NWFET without nanogap cavity and interface charges

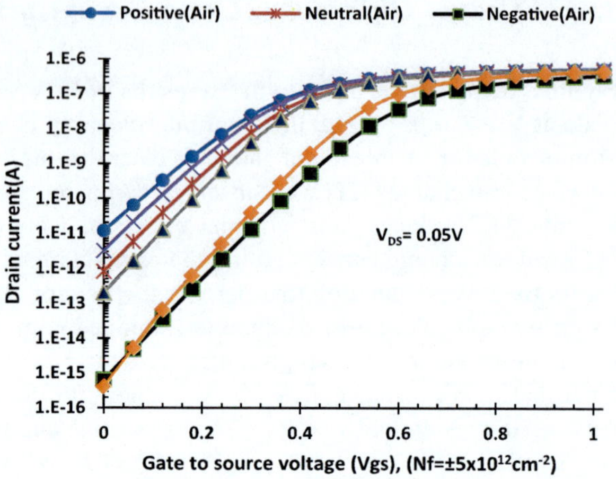

Fig. 3 Impact of interface trap charges on I_D-V_G, with air and biotin biomolecule for n-type triple metal gate high-k gate all around junctionless NWFET

3.2 *Impact of Interface Trap Charges on Switching Ratio and Leakage Current*

Figure 4 illustrates the variation of (a) switching ratio (b) leakage current, with different interface trap charges for n-type triple metal gate high-k GAA-JL-NWFET at $V_{DS} = 0.05$ V and $V_{GS} = 1.0$ V with air and biotin biomolecules. It is clear that higher leakage current and lower switching ratios are examined for air compared to biotin biomolecule. The reason is that the lower gate dielectric constant raises gate parasitic resistance and short channel effects. At higher dielectric constant (biotin) biomolecule, higher switching ratio and lower leakage current are explored with negative interface trap charges (ITCs) compared to positive and neutral ITCs due to an increasing number of mobility carriers across the channel by negative ITCs, thereby reducing DIBL [16, 17].

3.3 *Impact of Interface Trap Charges (ITCs) on Total Current Density*

Figure 5 demonstrates higher variation of total current density in the case of biotin biomolecule compared to air in the presence of ITCs. Example, ($N_f = \pm 5 \times 10^{12}\ \text{cm}^{-2}$) total current density for biotin ($\epsilon = 2.63$) are $8.44 \times 10^5\ \text{A/cm}^2$ and $2.39 \times 10^5\ \text{A/cm}^2$ for positive and negative ITCs, respectively, and air ($\epsilon = 1.0$) are $6.06 \times 10^5\ \text{A/cm}^2$ and $1.32 \times 10^5\ \text{A/cm}^2$ for positive and negative ITCs, respectively. The improved total current density of biotin biomolecule in comparison to air is 39.3% and 81% for positive and negative ITCs, respectively. This variation of overall current density indicates that our proposed device can detect biomolecules.

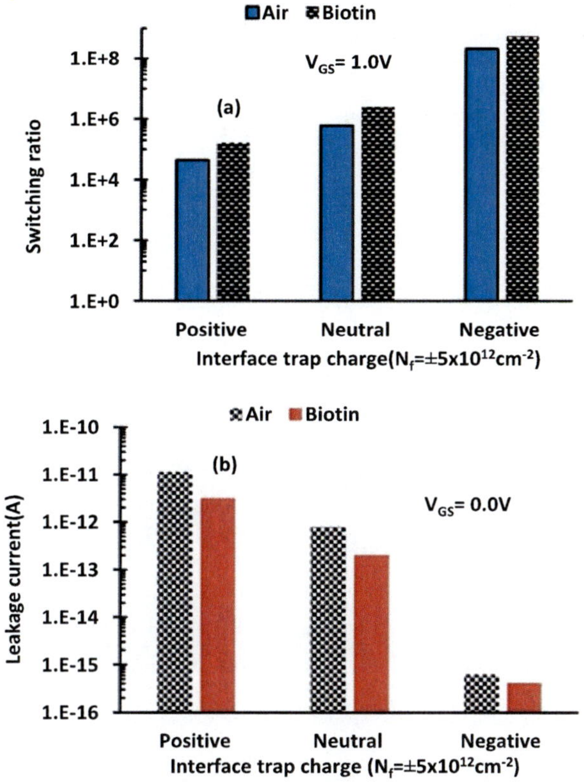

Fig. 4 Variation of **a** switching ratio **b** leakage current, with interface trap charges for n-type triple metal gate high-k GAA-junctionless NWFET

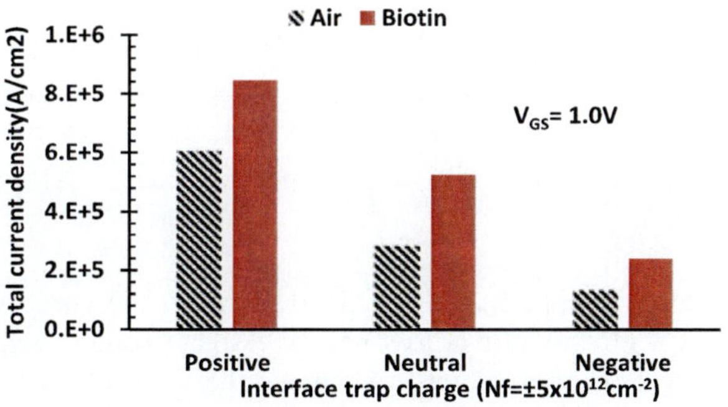

Fig. 5 Variation of total current density at drain end with ITCs for n-type triple metal gate high-k GAA-JL-NWFET

3.4 Effect of Interface Trap Charges (ITCs) on Sensitivity ($S_{I_{OFF}}$)

Figure 6 illustrates a higher drain-off current ratio ($S_{I_{off}}$) for positive and neutral than negative interface trap charges (ITCs); this is the result of higher leakage current due

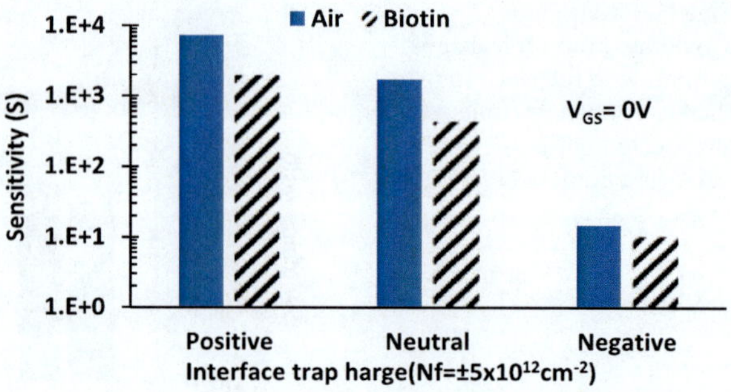

Fig. 6 Variation of drain-off current ratio with interface trap charges for n-type triple metal gate high-k GAA-JL-NWFET

to positive and neutral interface trap charges (TCs) led to higher drain-off current ratio ($S_{I_{off}}$) than negative ITCs; because of SCEs and DIBL effect experienced by positive and neutral interface trap charges (ITCs). For instance, at higher dielectric constant (biotin), drain of current ratio is smaller than that of lower dielectric constant (air); this indicates that high dielectric materials enhance gate electrostatic control, thereby reduces gate tunneling current. Sensitivity ($\mathbf{S_{I_{off}}}$) or drain-off current ratio is given by Eq. (1)

$$S_{I_{off}} = \frac{I_{off}\ (\text{with biomolecule Species})}{I_{off}\ (\text{without biomolecule Species})}\bigg|\text{at}V_{gs} = 0 \quad (1)$$

3.5 *Impact of ITCs on the Device Transconductance*

Transconductance (g_m) of a MOSFET characterizes its analog performance, and it is given by Eq. (2) [19] at constant drain-source voltage.

$$g_m = 2\left(\frac{I_{D\,ON}}{(V_{GS\,ON} - V_{th})^2}\right)V_{eff} \quad (2)$$

where (V_{GSON}, I_{DON}) is a fixed point on the curve when the device is ON, and V_{eff} is effective (overdrive) or excess gate voltage, which is the difference between the gate to source bias voltage and the threshold voltage, i.e., ($V_{eff} = V_{GS} - V_{th}$). Transconductance ($g_m$), also known as mutual conductance or transfer admittance, is a property of certain electronic components and is used to analyze MOSFET amplifiers.

As shown in Fig. 7b, transconductance for negative interface trap charge (ITCs) is higher than positive ITC, due to reduced hot carrier and DIBL effects in the presence of negative ITCs at strong inversion region. Example, at $V_{GS} = 1.0$ V and $V_{DS} = 0.6$ V for $N_f = \pm 5 \times 10^{12} cm^{-2}$ of ITCs, 20.3, and 33.7% are enhanced/raised transconductance of the device for negative and positive ITCs when

Fig. 7 Variation of **a** transconductance **b** transconductance at $V_{DS} = 0.6$ V with ITCs for n-type triple metal gate high-k gate all around JL-NWFET

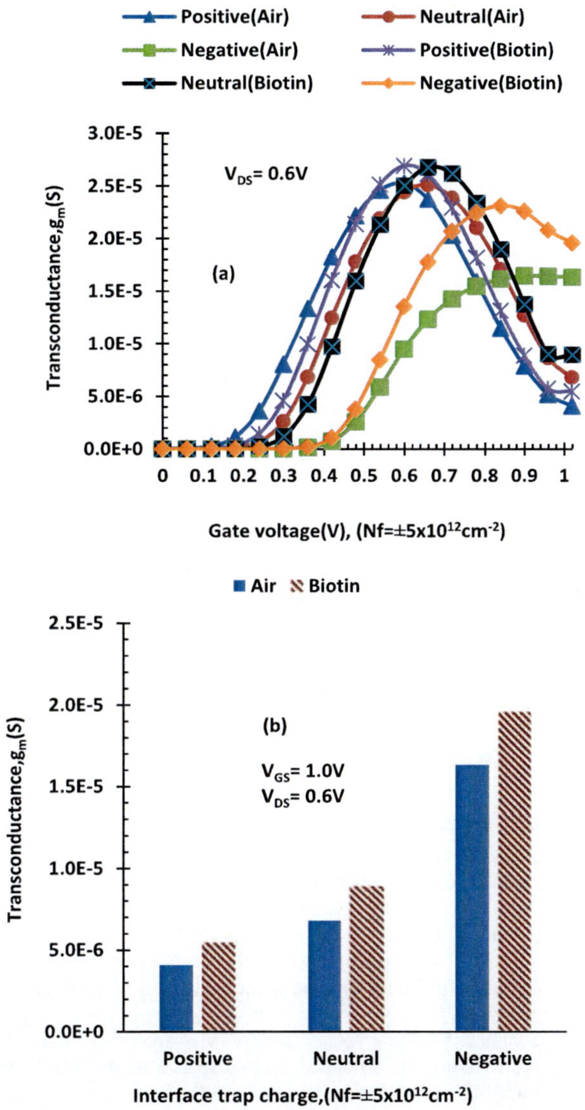

biotin biomolecule immobilizes/interacts with the nanogap cavity region compared to the air cavity. So that biotin biomolecule at negative interface trap charges delivers higher device gain (amplification) and more drive current experience lower SCEs [20] due to increased carrier injection by negative (interface trap charges) ITC across the channel.

4 Conclusion

In this study, the effect of different interface trap charges (ITCs) on transfer characteristics of TG-high-k-GAA-JL-NWFET-based biotin biosensor has been examined. These different ITCs cause the change in total current density, switching ratio, transconductance, leakage current, and drain-off current ratio when air and biotin biomolecule immobilize interchangeably on the nanogap region. In this study, drain-off current ratio is taken as a sensing metric to study the impact of ITCs on device sensitivity. In our research, for biotin biomolecule, enhanced drain-off current ratio (S) is 72.9% for positive and 29.7% for negative ITCs compared to air. The result indicates that biotin biomolecule shows a strong dependency on negative ITCs for increasing output parameters like switching ratio, transconductance, drain-on current, and total current density. Finally, we have concluded, negative ITC has a positive impact on our proposed device performance compared to positive ITC. In our study, we have examined that biotin biomarker for the silicon-based device presents reactive amine groups on the silicon surface for detecting cardiovascular diseases.

Acknowledgements This study is supported by Debre Tabor University, Ministry of Science and Higher Education of Ethiopia and Delhi Technological University, Delhi, India.

References

1. Rony, M.W., Bhowmik, P., Myler, H.R., Mondol, P.: Short channel effects suppression in a dual-gate gate-all-around Si nanowire junctionless nMOSFET. Proc. 9th Int. Conf. Electr. Comput. Eng. ICECE 2016 538–541 (2017)
2. Input, M.S.: Impact of hot carrier degradation on MOSFET small-signal input, output, and transmission features. In: ICCDCS (2014)
3. Ghosh, S., Chattopadhyay, A., Tewari, S.: Optimization of hetero-gate-dielectric tunnel FET for label-free detection and identification of biomolecules. IEEE Trans. Electron Devices **67**, 1–8 (2019)
4. Swain, S.K., Das, S.K., Biswal, S.M., Adak, S., Nanda, U., Sahoo, A.A., Navak, D., Baral, B., Tripathy, D.: Effect of high-K spacer on the performance of non-uniformly doped DG-MOSFET. Proc. 3rd Int. Conf. 2019 Devices Integr. Circuit, DevIC 510–514 (2019)
5. Kumar, V., Gupta, R., Preet, R., Singh, P., Vaid, R.: Performance analysis of double gate n-FinFET using high-k dielectric materials. Int. J. Innov. Res. Sci. Eng. Technol. **5**, 13242–13249 (2016)
6. Thapa, N., Maurya, L.: Performance advancement of High-K dielectric MOSFET. Int. J. Innov. Adv. Comput. Sci. **3**, 98–103 (2014)
7. Ajay, Narang, R., Saxena, M., Gupta, M.: Investigation of dielectric modulated (DM) double gate (DG) junctionless MOSFETs for application as a biosensors. Superlatt. Microstruct. **85**, 557–572 (2015)
8. Goel, A., Rewari, S., Verma, S., Gupta, R.S.: Dielectric modulated triple metal gate all around MOSFET (TMGAA) for DNA bio-molecule detection. 2018 IEEE Electron Devices Kolkata Conf. **1**, 337–340 (2019)

9. Madan, J., Chaujar, R.: Numerical simulation of N+ source pocket PIN-GAA-tunnel FET: impact of interface trap charges and temperature. IEEE Trans. Electron Devices **64**, 1482–1488 (2017)
10. Passeri, D., Morozzi, A., Kanxheri, K., Scorzoni, A.: Numerical simulation of ISFET structures for biosensing devices with TCAD tools. Biomed. Eng. Online **14**, S3 (2015)
11. Chhabra, A., Kumar, A., Chaujar, R.: Sub-20 nm GaAs junctionless FinFET for biosensing application. Vacuum **160**, 467–471 (2018)
12. Colinge, J.P., Lee, C.W., Afzalian, A., Akhavan, N.D., Yan, R., Ferain, I., Razavi, P., O'Neill, B., Blake, A., White, M., Kelleher, A.M., McCarthy, B., Murphy, R.: Nanowire transistors without junctions. Nat. Nanotechnol. **5**, 225–229 (2010)
13. Mohapatra, S.K., Pradhan, K.P., Sahu, P.K.: Influence of high-k gate dielectric on nanoscale DG-MOSFET. Int. J. Adv. Sci. Technol. **65**, 19–26 (2014)
14. Windbacher, T., Sverdlov, V., Selberherr, S.: Biotin-streptavidin sensitive BioFETs and their properties. Commun. Comput. Inf. Sci. **52**, 85–95 (2010)
15. Pirner, H.M.: Biotin sensing in Saccharomyces cerevisiae is mediated by a conserved DNA element and requires the activity of biotin-protein ligase. J. Biol. Chem. **281**, 12381–12389 (2006)
16. Lowe, B.M., Sun, K., Green, N.G.: Biosensing: sensitivity enhancement using streptavidin—biotin as a model system †. R. Soc. Chem. 4173–4200 (2017)
17. Saha, P., Banerjee, P., Dash, D.K., Sarkar, S.K.: Impact of trapped interface charges on short channel characteristics of WFE high-K SOI MOSFET. Proc. 3rd Int. Conf. 2019 Devices Integr. Circuit, DevIC 2019 118–123 (2019)
18. Chandan, B.V., Nigam, K., Sharma, D., Pandey, S.: Impact of interface trap charges on doping-less tunnel FET for enhancement of linearity characteristics. Appl. Phys. A Mater. Sci. Process. **124**, 1–6 (2018)
19. Cartwirght, K.V.: Derivation of the exact transconductance of a FET without calculus. Technol. Interface J. **10**, 7 (2009)
20. Yuan, J., Woo, J.C.S.: A novel split-gate MOSFET design realized by a fully silicided gate process for the improvement of transconductance and output resistance. IEEE Electron Device Lett. **26**, 829–831 (2005)

Fin Aspect Ratio Optimization of Novel Junctionless Gate Stack Gate All Around (GS-GAA) FinFET for Analog/RF Applications

Bhavya Kumar and Rishu Chaujar

1 Introduction

To keep up with Moore's statement, the size of the transistor reduces from the micrometer regime to the nanometer regime and is now reaching its fundamental limits. The performance of the device increases relatively with the scaling, but it also presents severe Short-Channel Effects (SCEs), which reduces the efficiency of the device [1]. Furthermore, to form an ultra-sharp S/D junction at such small dimensions is not an easy task. Colinge et al. proposed the Junctionless Field-Effect Transistors (JLFETs), which ease the fabrication process and lower the fabrication budget of the device [2]. Over the years, several multi-gate (MG) devices have been introduced by the researchers to alleviate the SCEs. FinFET, a "3D" field-effect transistor, appeared as the most encouraging device in comparison to other MG devices. Gate All Around (GAA) structure was proposed further to improve the subthreshold characteristics and performance of the FinFET. The ITRS also termed the GAA configuration as the "ultimate structure" [3, 4]. A decrease in the gate oxide thickness below a particular value enhances the gate direct tunneling current [5]. The most technological solution is the use of high-κ dielectric materials, like HfO_2 ($\kappa = 25$) [6]. However, there are issues with the direct deposition of high-κ dielectrics on Si substrates like mobility degradation and threshold voltage instability [7, 8]. The solution to this problem is the use of Gate Stack (GS) configuration comprising of a thin interfacial SiO_2 layer as passivation between the Si substrate and high-κ dielectrics by maintaining the effective oxide thickness constant [9]. Thus, we are proposing a novel Junctionless Gate Stack Gate All Around (GS-GAA) FinFET. TiN metal gate is preferred over

B. Kumar · R. Chaujar (✉)
Department of Applied Physics, Delhi Technological University, Delhi, India
e-mail: chaujar.rishu@dtu.ac.in

B. Kumar
e-mail: bhavyakumar_phd2k18@dtu.ac.in

A. Biswas et al. (eds.), *Microelectronics, Circuits and Systems*, Lecture Notes in Electrical Engineering 755, https://doi.org/10.1007/978-981-16-1570-2_6

the polysilicon gate due to its compatibility with CMOS processing, low resistivity, and thermal stability [10].

It has already been demonstrated by the researchers in the published work that the performance of a FinFET device mainly hangs on the variation of geometrical parameters like the height of Fin (H_{Fin}), the width of the Fin (W_{Fin}), and gate length (L_g) [11, 12]. In some research papers, Fin AR is described as the ratio of the W_{Fin} to the H_{Fin}, but the correct definition of Fin AR is the ratio of H_{Fin}/W_{Fin}. It is because if $W_{Fin} >> H_{Fin}$, the top surface orientation becomes more dominant for device characteristics, and the device acts as a planar device whereas if $W_{Fin} <= H_{Fin}$, the sidewall surface becomes more prevalent compared to the top surface and consequently the device acts as a 3D device. In the studies performed on the Fin aspect ratio variation, both H_{Fin} and W_{Fin} were altered without fixing the effective channel area ($H_{Fin} \times W_{Fin}$) of the device to a constant value. In this scenario, the conductance of the channel will change, and we will not get the appropriate impact of the Fin aspect ratio variation on the evaluated parameters [13]. Therefore, in this work, the effective channel area of the device is kept constant while varying the Fin aspect ratio and essential electrostatic, analog, and RF parameters are evaluated for analog/RF applications.

2 Device Structure and Simulation Methodology

Figure 1a displays the proposed 3D structure of Junctionless GS-GAA FinFET, whereas Fig. 1b and 1c portrays the vertical and horizontal cross-sectional view cut along the silicon fin of the proposed device, respectively. The gate length (L_g) and the length of source/drain regions ($L_{S/D}$) are fixed at 7 nm and 10 nm, respectively. The oxide thickness (t_{ox}) is set at 1 nm, and SiO_2 ($\kappa = 3.9$) and HfO_2 ($\kappa = 25$) have been used in equal proportions for the stacking of the gate oxide. All the regions

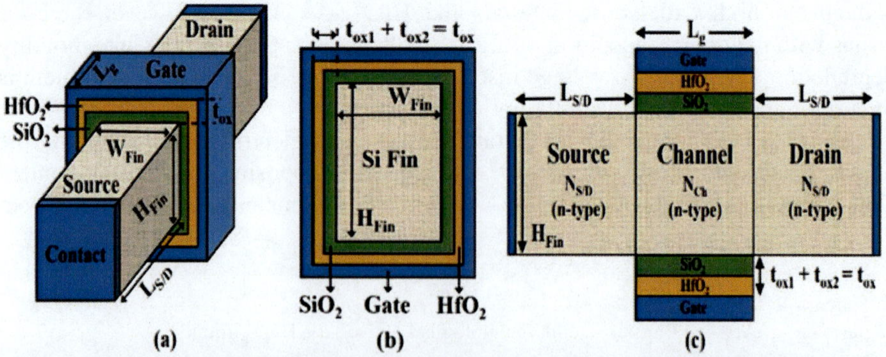

Fig. 1 **a** 3D structure of proposed Junctionless GS-GAA FinFET, **b** vertical and **c** horizontal cross-sectional view cut along the silicon Fin of Junctionless GS-GAA FinFET

Table 1 Different configurations used for simulation

Config. No	Fin height, H_{Fin} (nm)	Fin width, W_{Fin} (nm)	Channel area $H_{Fin} \times W_{Fin}$ (nm^2)	Fin aspect ratio, AR (H_{Fin}/W_{Fin})
C1	10	8	80	1.25
C2	16	5	80	3.2
C3	20	4	80	5

are uniformly doped with n-type doping species to make it a junctionless device. The doping concentration of channel (N_{Ch}) and source/drain regions ($N_{S/D}$) is 1×10^{19} cm^{-3}. The work function (ϕ_G) of the TiN metal gate is 4.65 eV [14]. In the entire fin region, silicon (Si) material is used with height (H_{Fin}), and width (W_{Fin}) of the fin varied in such a way that the effective channel area ($H_{Fin} \times W_{Fin}$) remains constant at 80 nm^2. We have considered three different configurations for simulation, namely, C1, C2, and C3 placed in increasing Fin aspect ratio order in Table 1. All the three different designs have been simulated using the SILVACO ATLAS 3D simulator [15]. During the simulations, V_{gs} changes from 0 to 1 V, V_{ds} is fixed at 0.5 V, and the temperature is 300 K. The threshold voltage (V_{th}) is used to evaluate the gate overdrive voltage (V_{gt}), which is the difference between gate-source voltage and threshold voltage, as displayed in Eq. (1). In amplifier circuits, the region of operation is decided by V_{gt}. Thus, the evaluated parameters are analyzed against V_{gt}.

$$V_{gt} = V_{gs} - V_{th} \tag{1}$$

Several physical models are included in device simulation to obtain more realistic and accurate results. The Arora model is invoked to incorporate concentration-dependent mobility. The accuracy in the results is increased with the help of the Fermi–Dirac statistics model. SRH model is employed for recombination and generation effects. Klaassen band-to-band tunneling model is included to account for the tunneling of electrons, and the Crowell–Sze impact ionization model is enabled to consider the impact ionization effects. Further, Gummel and Newton's methods are used to attain a solution [15]. The physical models have been validated with the published results of Hyunjin Lee et al. [16]. Figure 2a outlines the experimental and simulated transfer characteristics of a 5 nm All Around Gate (AAG) FinFET at V_{ds} = 1.0 V. The results are well-calibrated, thus validating the choice of simulation models used in the device simulation.

3 Results and Discussions

Figure 3a and 3b represents the transfer characteristics (I_d–V_{gs}) at a fixed drain–source voltage (V_{ds}) = 0.5 V in linear and log scale for different configurations. In Fig. 3a, it is observed that drain current (I_d) increases with the increase in the

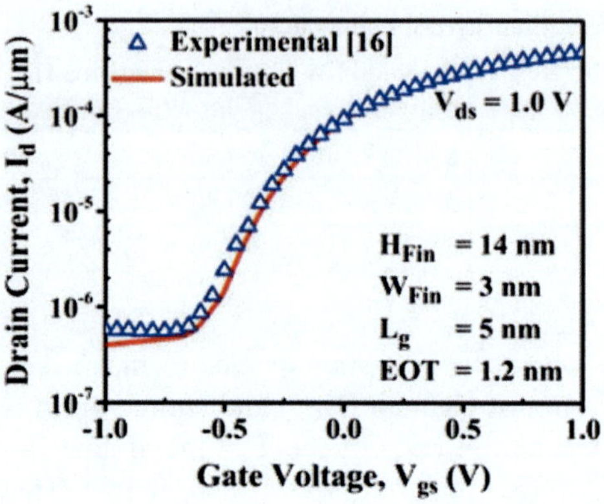

Fig. 2 Calibration of simulated and experimental transfer characteristics of a 5 nm All Around Gate (AAG) FinFET

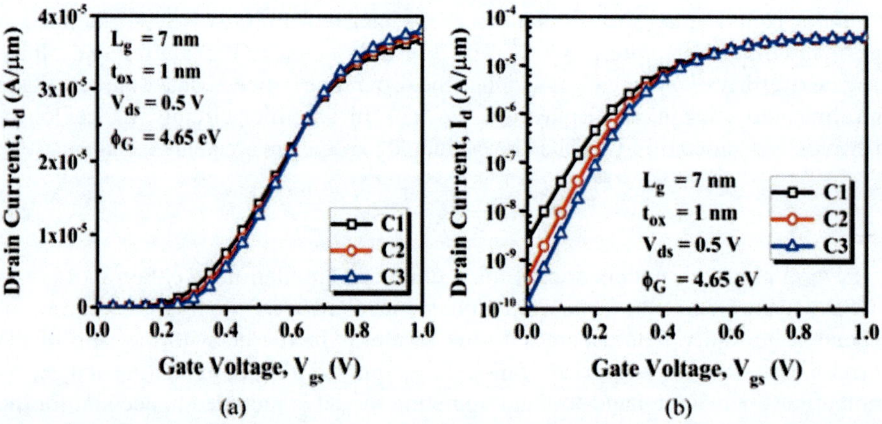

Fig. 3 Variation of transfer characteristics in **a** linear scale and **b** log scale for different configurations

Fin aspect ratio. As the Fin aspect ratio increases, current drivability increases, and consequently, C3 configuration acquires the maximum value of I_d in comparison to two other configurations. Also, as displayed in Fig. 3b, leakage current (I_{off}) improves as the Fin AR increases. I_{off} reduces approximately by 20 times for C3 configuration in comparison to C1 configuration. The reason is that as the Fin AR increases (either fin gets taller or narrower or both), there is a reduction in the electric field in the Fin region, which in turn reduces the leakage current [12].

In Fig. 4a, switching ratio (I_{on}/I_{off}) is plotted for each configuration. A higher switching ratio (I_{on}/I_{off}) is obtained for C3 configuration owing to increased drain current (I_d) and reduced leakage current (I_{off}), as shown in Fig. 3a and 3b, respectively. The I_{on}/I_{off} ratio increases by 19.58 times for C3 configuration in comparison

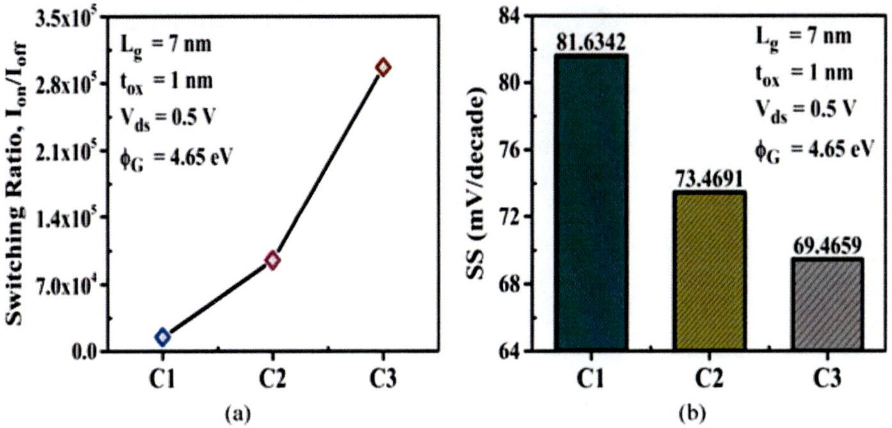

Fig. 4 Comparison of **a** switching (I_{on}/I_{off}) ratio and **b** subthreshold swing (SS) for different configurations

to the C1 configuration. Figure 4b depicts the subthreshold swing (SS) for all three configurations. Subthreshold swing (SS) is a crucial parameter to measure the Short-Channel Effects (SCEs) in the microscopic device. It is visible from Fig. 4b that for higher Fin aspect ratio, SS reduces and is nearest to its ideal value of 60 mV/decade. For C3 configuration, SS reduces by 1.17 times in comparison to the C1 configuration.

In analog applications, transconductance (g_m) and device efficiency (TGF) are the most critical performance parameters. Transconductance, as defined in Eq. (2), computes the change in I_d to the shift in V_{gs} at constant V_{ds}. In Fig. 5a, g_m is plotted as a function of V_{gt} for different Fin aspect ratio configurations. It is observed that

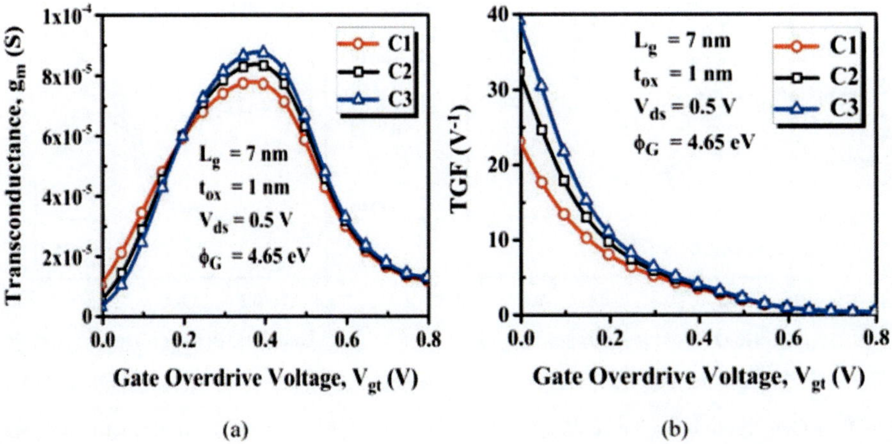

Fig. 5 **a** Transconductance (g_m) and **b** device efficiency (TGF) as a function of gate overdrive voltage (V_{gt}) for different configurations

g_m increases as the Fin AR increases with C3 configuration providing 1.13 times higher transconductance in comparison to C1 configuration. Figure 5b represents the variation of device efficiency (TGF) with V_{gt} for each Fin AR configuration. TGF is defined as the amount of gain generated per unit power loss, as expressed in Eq. (3). In Fig. 5b, the variation in the TGF is observed only in the subthreshold region with almost no change in the strong inversion region. TGF increases with the Fin aspect ratio because the higher value of drain current corresponds to higher g_m, and consequently higher TGF is recorded for C3 configuration.

$$g_m = \partial I_d / \partial V_{gs} \tag{2}$$

$$TGF = g_m / I_d \tag{3}$$

Cutoff frequency (f_T) and maximum oscillation frequency (f_{max}) are the two most essential parameters from the RF applications point of view, and to evaluate them, we need to extract the value of gate–drain capacitance (C_{gd}) and total gate capacitance (C_{gg}). The values of C_{gd} and C_{gg} have been extracted using the small-signal AC analysis at an operating frequency of 1 MHz with DC voltage ramped from 0 to 1 V while maintaining a step size of 0.05 V. In Fig. 6a and 6b, gate–drain capacitance and total gate capacitance are displayed as a function of V_{gt} for all three configurations. In comparison to C1 configuration, a higher value of C_{gd} and C_{gg} is obtained for the C3 configuration. The fundamental reason is that as the Fin aspect ratio increases, i.e., either fin gets taller or narrower or both, the fringing field in the device increases, which in turn enhances the capacitance in the device. Thus, an increase in the Fin aspect ratio affects the performance of the device.

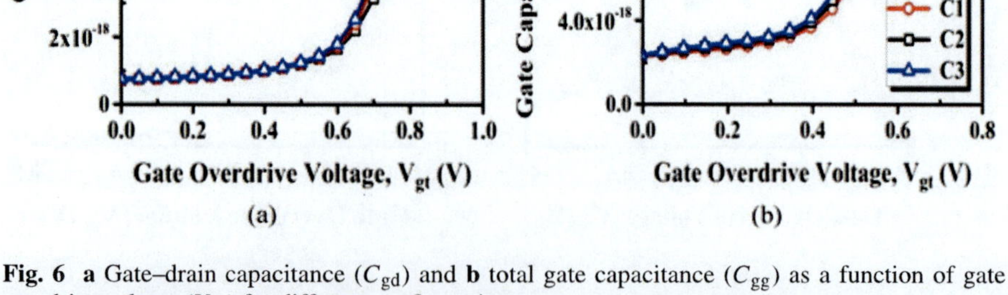

Fig. 6 **a** Gate–drain capacitance (C_{gd}) and **b** total gate capacitance (C_{gg}) as a function of gate overdrive voltage (V_{gt}) for different configurations

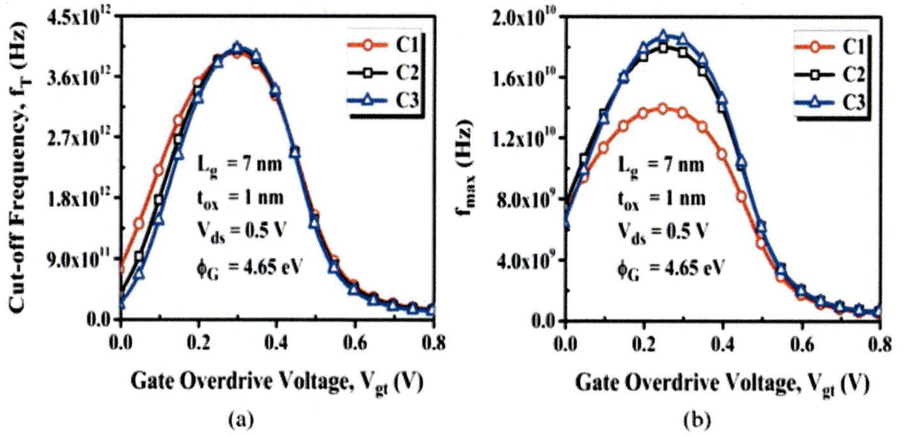

Fig. 7 **a** Cutoff frequency (f_{T}) and **b** maximum oscillation frequency ($f_{\max}$) as a function of gate overdrive voltage (V_{gt}) for different configurations

The cutoff frequency (f_{T}), as defined in Eq. (4), is the frequency at which current gain becomes unity (0 dB). Figure 7a outlines the plot of cutoff frequency against gate overdrive voltage (V_{gt}). It is observed that f_{T} increases with the Fin aspect ratio. A slightly greater value of cutoff frequency is obtained for C3 configuration in comparison to the other two configurations. The rise in f_{T} is minimal because of the increased value of total gate capacitance (C_{gg}), which suppresses the enhancement in transconductance (g_{m}) [17]:

$$f_T = g_m/2\pi\left(C_{\mathrm{gs}} + C_{\mathrm{gd}}\right) \tag{4}$$

$$f_{\max} = f_T/\surd\left\{4R_g\left(g_{\mathrm{ds}} + 2\pi\ f_T C_{\mathrm{gd}}\right)\right\} \tag{5}$$

The maximum oscillation frequency ($f_{\max}$) is the frequency at which maximum unilateral power gain becomes unity (0 dB) [18]. The maximum oscillation frequency ($f_{\max}$) is evaluated using Eq. (5) in which R_{g}, g_{ds}, and C_{gd} represent the gate resistance, drain–source output conductance, and gate–drain capacitance, respectively. The value of R_{g} is substituted as 50 Ω, whereas g_{ds} and C_{gd} are extracted using the small-signal AC analysis. Figure 7b exhibits the plot of maximum oscillation frequency ($f_{\max}$) as a function of V_{gt} for all three configurations. It is observed from Fig. 7b that $f_{\max}$ increases with the increase in Fin aspect ratio and 1.33 times higher values of $f_{\max}$ are obtained for C3 configuration in comparison to C1 configuration. It is because, in our work, the cutoff frequency is almost the same, but drain–source output conductance (g_{ds}) reduces significantly for higher Fin AR, which in turn enhances the value of maximum oscillation frequency ($f_{\max}$).

4 Conclusions

In this paper, we investigated the impact of Fin AR on Junctionless GS-GAA FinFET in terms of analog and RF parameters using the SILVACO ATLAS 3D simulator. The result indicates that for the C3 configuration (i.e., AR = 5), the switching ratio increased immensely by 1858.18% when compared to C1 configuration (i.e., AR = 1.25). In contrast, leakage current (I_{off}) and subthreshold swing (SS) get reduced by 94.71% and 14.90%, respectively. Therefore, the high Fin aspect ratio enhances the device performance and suppresses the SCEs as well. Furthermore, parameters like TGF, f_{max}, g_{m}, and f_{T} exhibit a significant improvement as well for C3 configuration with TGF enhanced by 69.9%, f_{max} by 33.8%, g_{m} by 13.07%, and f_{T} by 1.52%. Thus, for the designing of analog and RF circuits, junctionless GS-GAA FinFET with a high Fin aspect ratio can be looked upon as a tempting solution. However, we also have to consider the fabrication limitations in creating such narrow Fins with a high aspect ratio as the chances of the fin falling over also increases.

References

1. Kumar, A., Gupta, N., Chaujar, R.: TCAD RF performance investigation of transparent gate recessed channel MOSFET. Microelectron. J. **49**, 36–42 (2016)
2. Colinge, J.P., Lee, C.W., Afzalian, A., Akhavan, N.D., Yan, R., Ferain, I., Razavi, P., O'Neill, B., Blake, A., White, M., Kelleher, A.M., McCarthy, B., Murphy, R.: Nanowire transistors without junctions. Nat. Nanotechnol. **5**(3), 225–229 (2010)
3. The International Technology Roadmap for Semiconductors. https://www.itrs2.net/.
4. Kumar, B., Kumar, A., Chaujar, R.: The effect of gate stack and high-κ spacer on device performance of a junctionless GAA FinFET. In: 2020 IEEE VLSI Device Circuit and System (VLSI DCS), pp. 159–163. IEEE, Kolkata (2020)
5. Lo, S.H., Buchanan, D.A., Taur, Y., Wang, W.: Quantum-mechanical modeling of electron tunneling current from the inversion layer of ultra-thin-oxide nMOSFET's. IEEE Electron. Device Lett. **18**(5), 209–211 (1997)
6. Kumar, A., Gupta, N., Tripathi, S.K., Tripathi, M.M., Chaujar, R.: Performance evaluation of linearity and intermodulation distortion of nanoscale GaN-SOI FinFET for RFIC design. AEU Int. J. Electron. Commun. **115**, 153052 (2020)
7. Kerber, A., Cartier, E., Pantisano, L., Degraeve, R., Kauerauf, T., Kim, Y., Hou, A., Groeseneken, G., Maes, H.E., Schwalke, U.: Origin of the threshold voltage instability in SiO_2/HfO_2 dual layer gate dielectrics. IEEE Electron. Dev. Lett. **24**(2), 87–89 (2003)
8. Onishi, K., Kang, C.S., Choi, R., Cho, H.J., Gopalan, S., Nieh, R.E., Krishnan, S.A., Lee, J.C.: Improvement of surface carrier mobility of HfO_2 MOSFETs by high-temperature forming gas annealing. IEEE Trans. Electron. Dev. **50**(2), 384–390 (2003)
9. Gupta, N., Chaujar, R.: Optimization of high-k and gate metal workfunction for improved analog and intermodulation performance of Gate Stack (GS)-GEWE-SiNW MOSFET. Superlattices Microstruct. **97**, 630–641 (2016)
10. Liu, Y., Kijima, S., Sugimata, E., Masahara, M., Endo, K., Matasukawa, T., Ishii, K., Sakamoto, K., Sekigawa, T., Yamauchi, H., Takanashi, Y., Suzuki, E.: Investigation of the TiN gate electrode with tunable work function and its application for FinFET fabrication. IEEE Trans. Nanotechnol. **5**(6), 723–728 (2006)
11. Kranti, A., Raskin, J.P., Armstrong, G.A.: Optimizing FinFET geometry and parasitics for RF applications. In: International SOI Conference Proceedings, pp. 123–124. IEEE (2008)

12. Mohapatra, S.K., Pradhan, K.P., Singh, D., Sahu, P.K.: The role of geometry parameters and Fin aspect ratio of sub-20nm SOI-FinFET: an analysis towards analog and RF circuit design. IEEE Trans. Nanotechnol. **14**(3), 546–554 (2015)
13. Biswas, K., Sarkar, C.K.: Optimizing Fin aspect ratio of junctionless bulk FinFET for application in Analog/RF circuit. In: Electron Devices Kolkata Conference (EDKCON), pp. 591–595. IEEE, Kolkata (2018)
14. Vitale, S.A., Kedzierski, J., Healey, P., Wyatt, P.W., Keast, C.L.: Work-function-tuned TiN metal gate FDSOI transistors for subthreshold operation. IEEE Trans. Electron. Dev. **58**(2), 419–426 (2011)
15. SILVACO, I.: ATLAS User's Manual. Santa Clara, CA, Ver 5 (2016)
16. Lee, H., Yu, L.E., Ryu, S.W., Han, J.W., Jeon, K., Jang, D.Y., Kim, K.H., Lee, J., Kim, J.H., Jeon, S.C., Oh, J.S., Park, Y.C., Bae, W.H., Lee, H.M., Yang, J.M., Yoo, J.J., Kim, S.I., Choi, Y.K.: Sub-5nm all-around gate FinFET for ultimate scaling. In: Digest of Technical Papers—Symposium on VLSI Technology, pp. 58–59 (2006)
17. Kumar, A., Tripathi, M., Chaujar, R.: Investigation of parasitic capacitances of In 2 O 5 Sn gate electrode recessed channel MOSFET for ULSI switching applications. Microsyst. Technol. **23**(12), 5867–5874 (2017)
18. Gupta, N., Kumar, A., Chaujar, R.: Impact of device parameter variation on RF performance of gate electrode workfunction engineered (GEWE)-silicon nanowire (SiNW) MOSFET. J. Comput. Electron. **14**(3), 798–810 (2015)

Design and Analysis of Cyl GAA-TFET-Based Cross-Coupled Voltage Doubler Circuit

Ankur Beohar, Ambika Prasad Shah, Nandakishor Yadav, Gopal Raut, and Santosh Kumar Vishvakarma

1 Introduction

1.1 Device Circuit Literature

The recent advancement in Internet of Thing (IoT) era promotes low yield power design, high driving capability, process optimization, and smart peripherals design with a reliability approach within devices, circuits to architecture levels. In particular, such designs are best suitable for utilization at a remote location for a longer period. Therefore, an IoT device required an energy harvesting circuit- or device-level techniques under a low range of voltage [1]. Further, energy harvesting comprises utilizing energy from surrounding and transforming it to electrical signal. In addition, the IoT system needs high speed with better energy efficiency design based on low yield power devices to fulfill the requirement of ultralow yield power applications [2].

Recently, many researchers have been specifically focusing on developing smart and small devices with their peripheral circuits to the external world for emerging IoT. This is because of low yield power sensor, which transmits the information toward destination is deployed at a remote location for IoT perspective [3, 4]. It also has an

A. Beohar (✉)
VIT Bhopal University, Bhopal, MP, India

A. P. Shah
Discipline of Electrical Engineering, Indian Institute of Technology Jammu, Jammu, J&K, India
e-mail: ambika.shah@iitjammu.ac.in

N. Yadav
Illinois Institute of Technology, Chicago, USA

G. Raut · S. K. Vishvakarma
Discipline of Electrical Engineering, Indian Institute of Technology Indore, Indore, MP, India
e-mail: skvishvakarma@iiti.ac.in

A. Biswas et al. (eds.), *Microelectronics, Circuits and Systems*, Lecture Notes in Electrical Engineering 755, https://doi.org/10.1007/978-981-16-1570-2_7

ever-increasing demand for longer battery life and low yield power emerging devices while maintaining high performance [5]. Therefore, addressing to the IoT system requirement, the power efficient emerging devices have been examined continuously under extreme short channel node to achieve Moore's law relating to the integrated circuit (IC) packaging density, efficient area utilization, and low power dissipation. Whereas assertive device technology node scaling produces appalling short channel effects (SCEs), which reflects to not only high leakage current but also significantly leads to high dynamic power consumption [6].

Addressing to this issue, the transistor with tunneling phenomena of electric field such as tunneling field-effect transistor (TFET) is the alternative and futuristic ideal device for on chip control system [7]. The property and beauty of the device is its sub-threshold swing and hence at room temperature it not stuck less than 60 mV/Dec. Thus, it produces low leakage/off current (I_{OFF}) as compared to the conventional field-effect transistor (FET) [8]. This property also contributes for the high barrier for the SCEs and low static power dissipation due to charge carrier tunneling phenomenon while moving from valence band to the conduction band of the tunneling device. Despite these benefits, device has some limitations such as ambipolar behavior and less ON current (I_{ON}) [9].

To overcome the limitation of the conventional architecture, the device is proposed having a cylindrical structure, i.e., gate all around (GAA) in shape across the axis. It causes fully inversion across the volume and effective channel control by the applied gate voltage. Recently, we demonstrated [10] a brief examination of three structures based on GAA-tunnel-FET. Previously, Kim et al. in [11] proposed low E_g material such as Ge-based source in n-tunnel-FET. The design based on Ge-source elucidates higher I_{ON} due to the low bandgap and high probability of tunneling rate. However, Ge-source-based structure produces high off-state current. Although its high I_{ON} deteriorates the performance. Thus, this work proposed Ge-source based on cyl GAA-tunnel-FET with low spacer width and investigated for high-performance and low yield power circuit. Here, low I_{OFF} and steepest *SS* achieved high drain-channel resistance. Note that the tunnel-FET in combination with *Si* and *Ge* material causes a superiority in their performances in such a way that tunnel-FET works under reverse biasing, and Si is the only material to provide the highest reverse saturation current during reverse biasing. In addition, it promotes ease of fabrication processing and lower cost due to its geometry of all-around symmetry across the axis.

The cross-coupled voltage doubler (CCVD) is the most suitable circuit for the requirement of applications that is used as energy harvesting circuits with IoT systems at remote locations. It can also be used to enhance the available low supply voltage up to the required supply voltage to operate the system, i.e., in switched capacitor especially a cross-coupled voltage doubler as it is one from the family. The output voltage is almost examined double of the input voltage. It also reduces the ripple in voltage with the same frequency as compared to the conventional charge pump [12]. This is mostly used with DC-to-DC converter over small battery source to power the IoT applications. However, the power efficiency of the CCVD is limited due to reversion loss. However, this topology also has a significant issue as drop in voltage due to the cutoff voltage of the diode. To address the issues, application to boost

DC-to-DC converter, the CMOS-based cross-coupled voltage multiplier is designed by Yu et al. in [13].

Furthermore, Kim et al. in [14] proposed the transfer blocking technique, which has no reversion loss. However, all of the above designs are based on conventional MOSFET technology that reflects an issue of high power consumption. Therefore, low yield energy device plays a vital role in powering such systems having limited energy capacity of batteries or to increase the lifetime of batteries. In this paper, we have also designed a CCVD circuit based on the proposed device with Ge-source and extreme low spacer width at V_{DS} of 0.35 V. This cross-coupled DC–DC converter reduces the reversion leakage loss and increases the power efficiency. It not only reduces the significant power loss but also increases the conversion efficiency along with minimal energy consumption.

2 Device and Simulation Models

GAA-based structures produces superior gate control over the inversion layer and supports for entire volume inversion, which attributes for low leakage applications. The various model and geometrical parameters associated with the device are: channel area doping = 10^{16} cm^{-3}, p+ (doping of Source) = 10^{20} cm^{-3}, n+ (doping of drain) = 5×10^{18} cm^{-3}, oxide thickness (t_{ox}) = 2 nm, and gate work function = 4.85 eV. In this structure, S is the symbolic of Source region while D represents Drain, and C stands for Channel of the device. Whereas gate length (L_G) = 25 nm, (L_{sw}) = 12 nm, R = 5 nm, (t_{ox}) = 2.5 nm drain extension length L_D = 40 nm, source extension length L_S = 40 nm, and L_{ch} = 20 nm are used. Further, supporting device model libraries are available for simulation along with band-to-band tunneling (BTBT). This models has a quantum phenomenon such as electrons tunneling from energy band-to-band transition within the device physics. The well-known technique of bandgap narrowing termed as Shockley–Read–Hall (SRH). It uses fixed minority carriers life time Fermi–Dirac. Moreover, it is applied to reduce carrier concentrations in heavily doped region. Here, AC analysis has also been performed under the frequency of 5 MHz. The schematic and three-dimensional view of underlap in drain (DU) of the proposed structure with low spacer width (LSW) is shown in Fig. 1a, b. The simulation results of transfer curve for drain current of n- and p-type Cyl GAA-tunnel-FET shown in Fig. 2. The device physical parameters considered are 25 nm gate length with drain voltage of 0.35–1.5 V.

The simulation result and device calibration with experimental data of [15] is shown in Fig. 3. Further, it has high I_{ON} (0.83×10^{-4} A/μm), low I_{OFF} (2.09×10^{-17} A/μm), and an enhanced I_{ON}/I_{OFF} (10^{13}) with which low SS of 25.8 mV/decade is achieved. This is because of placement of low spacer width over the source region consisting of Ge, which is lowering the fringing field effects and reduces tunneling barrier width at the source-gate edge. Thus, depletion zone not created results in the high source channel tunneling at the junction, not inside the body, which leads to high I_{ON}. Similarly, asymmetry in underlap increases the drain-channel barrier width

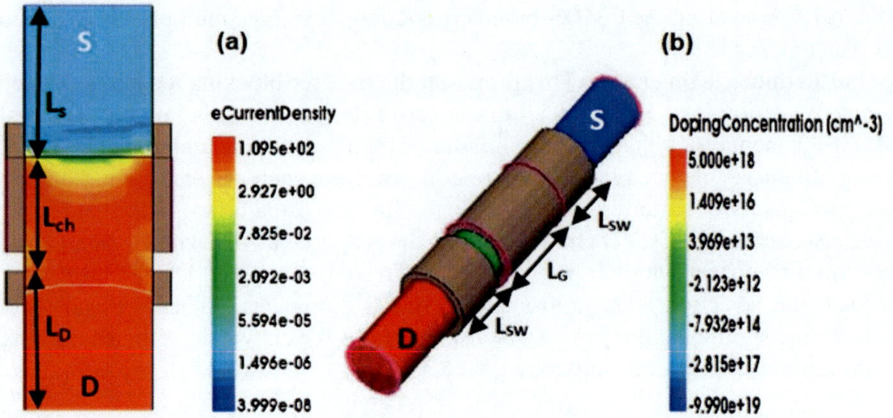

Fig. 1 **a** The simulated cyl-tunnel-FET structure cross-sectional view **b** shows the three-dimensional across the z-axis with asymmetric doping and parameters

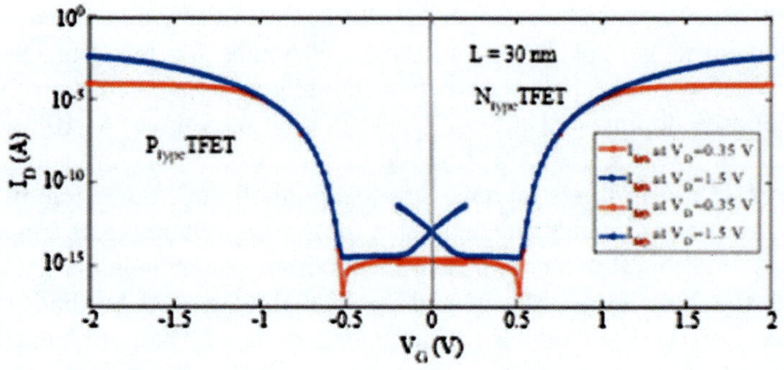

Fig. 2 I_D-V_G characteristics for the proposed device at V_D= 0.35 V and 1.5 V

and causes the electric field weakening, which does not affect I_{ON}. Consequently, it reduces the rate of tunneling at the drain-channel junction, and thus low I_{OFF} with steepest SS was achieved and compared with experimental data [16] as shown in Table 1.

Moreover, with reliability parameters [17], we exercise on a LUT (look up table) based on the various parameters with step size = 0.01 to form a spice model file tabulating $I_{ds}\$$ (v_{gs} and v_{ds}), $\$I_{gs}$ (v_{gs} and v_{ds}), c_{gs} (v_{gs} and v_{ds}), and c_{gd} (v_{gs} and v_{ds}) parameter properties of the proposed cyl GAA-tunnel-FET. With the help of Verilog-A device model file, circuit parameters were examined using circuit simulator tool. Figure 4a shows the inverter based on complementary tunnel-FET, respectively. Figure 4b demonstrates the simulated voltage transfer characteristics of inverter consisting of n-tunnel-FET and p-tunnel-FET based on proposed structure, which shows a similar characteristic as CMOS inverter. The results indicate that the

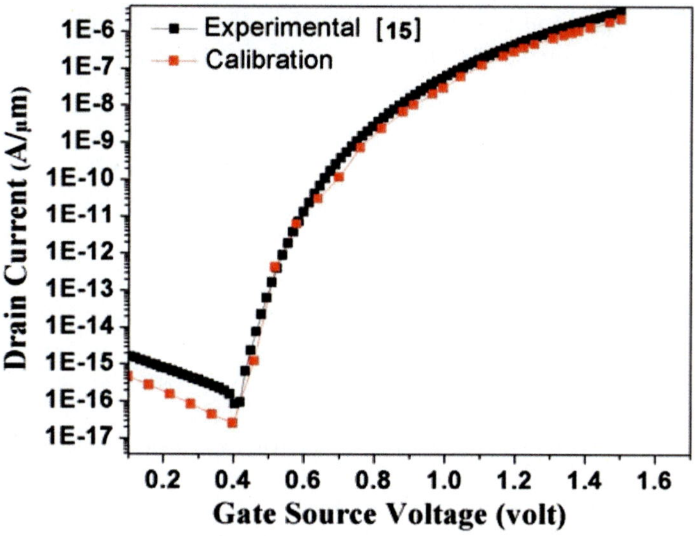

Fig. 3 Calibration of our simulation result with experimental data of [15]

Table 1 Device design parameters of different tunnel-FET structures

Performance parameters	This work		Moselund et al. [16]
	$V_{dd} = 0.35$ V	$V_{dd} = 1.5$ V	
I_{ON} (A/μm)	0.83×10^{-4}	3.8×10^{-4}	0.3×10^{-6}
I_{OFF} (A/μm)	2.09×10^{-17}	3.43×10^{-15}	9.2×10^{-13}
I_{ON}/I_{OFF}	0.39×10^{13}	1.1×10^{11}	1×10^{7}
SS (mV/dec)	25.8	28.3	90

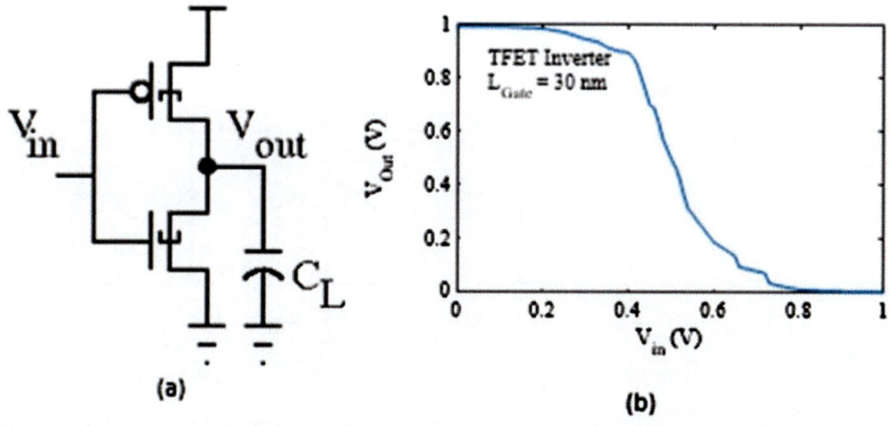

Fig. 4 Circuit performance **a** Cyl GAA-tunnel-FET-based inverter circuit with load capacitance (C_L). **b** Voltage transfer characteristics of CMOS inverter based on the proposed device

proposed device can be utilized to implement the circuits for low yield power applications. We extended our analysis by applying the CCVD using the proposed device, as discussed in the subsequent sections.

3 Cross-Coupled Voltage Doubler Based on Cyl GAA-Tunnel FET

With our proposed device of low bandgap material, we designed the CCVD circuit and investigated the performance. The CCVD based on the proposed structure produces superior transient and DC feature due to low depletion barrier width and no formation of depletion zones across source channel junction and increased depletion resistance for the drain and channel junction attributed as the cause of underlap in drain.

Figure 5a shows the circuit of CCVD based on Cyl GAA-tunnel-FET. In the circuit, T1 and T3 are the two transistors based on the proposed tunneling device forming the first inverter, whereas T2 and T4 form the second inverter for the cross-coupled operation. These two inverters are connected in cross-coupled mode, and storing nodes are connected with the charging capacitors. Here C_1 and C_2 are the two flying capacitors, and C_L is the load capacitor.

The switches Q1 and Q2, which can close and open the capacitors controlled by the two-phase non-overlapping clock, are shown in Fig. 5b. The operating frequency is set as 50 MHz for the non-overlapping clock signal. These clock signals must operate in the break-before-make fashion to reduce the shoot-through current during switching. In the steady state, the transistors are operated in the linear region.

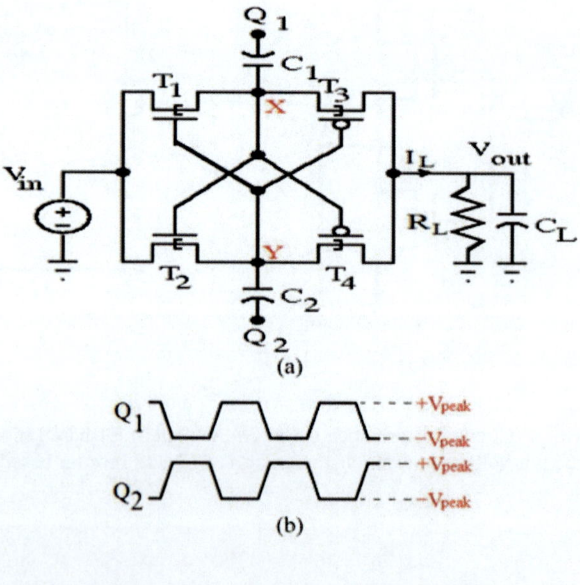

Fig. 5 Design of cross-coupled voltage doubler (CCVD) **a** schematic diagram and **b** control scheme of the proposed circuit

In the first phase, when switch Q_1 is ON, T_2 will ON, and T_4 will come under cutoff mode. Meanwhile, Q_2 is OFF, and charge voltage across C_2 causes the transistor T_3 to be ON. In this circuit, C_2 is charged by V_{in} voltage and Q_2 will also charge by V_{in} voltage through transistor T_2, and therefore C_2 voltage will be $V_{in} + V_{in}$. At the same time, T_3 is ON, and it passes output voltage through the C_1 capacitor to the R_L. Significantly, switch Q_2 is ON, it makes T_1 ON and T_3 OFF, leading to the charging of capacitor C_1 through additional voltage V_{in} and T_4 is ON through Q_1 that passes the C_2 charge to the output that is $2V_{in}$. The charging and discharging depend on the capacitance and ON currents of the transistors.

Further, in the proposed design, we can simplify the analysis for a two-phase non-cross triggering signal in order to ensure that the duty cycle should be below 50%. During the steady state for complete a clock cycle, the charges supplied by the power supply must be equal to the charges delivered to the load. Because the load current (I_L) is >0, the computation of charges provided in one clock cycle is given by $Q_T = I_L \times T$, where T is the time signal. The capacitors can be converted to the fly capacitors C_{FC1} and C_{FC2}. The driving current (I_D) of the output is near to I_L. Moreover, the current across the device has two parts, which are defined as capacitor charging current (I_C) and load current I_L.

4 Results and Discussions

The proposed CCVD based on GAA-tunnel-FET was simulated and implemented at 25 nm channel length. In this design, each of the flying capacitors C_1 and C_2 are 10 pF, and the load capacitor (C_L) is 1000 pF. The two auxiliary capacitors are 5 pF. The output voltage of the proposed design concerning time for the different values of input voltage while keeping the fixed value of load resistance (R_L) of 80 MΩ is shown in Fig. 6. From the results, we observe that the settling time decreases as the input voltage increases. The observed settling times are 150.4 ms, 72.9 ms, 60.5 ms,

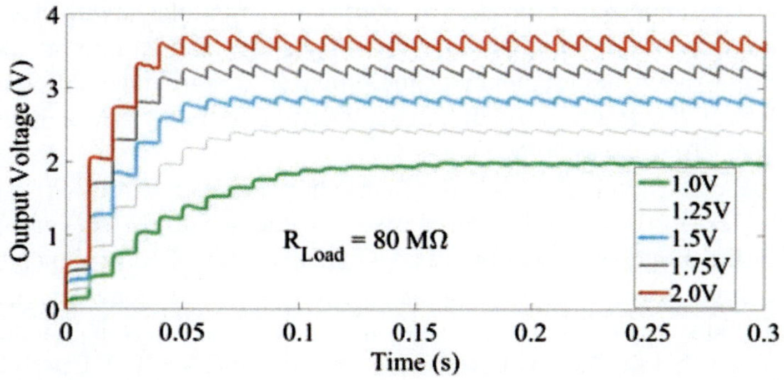

Fig. 6 Transfer characteristics for the output voltage of the CCVD design for five different values of input voltage (V_{in})

Table 2 The output voltage of CCVD based on MOSFET, FinFET and our work for different supply voltages

Input supply V_{dd} (V)	Output voltage (V)		
	MOSFET	FinFET	Tunnel-FET
1.00	1.90	1.91	1.98
1.25	2.35	2.40	2.65
1.50	2.56	2.75	2.89
1.75	3.10	3.27	3.28
2.00	3.53	3.40	3.89

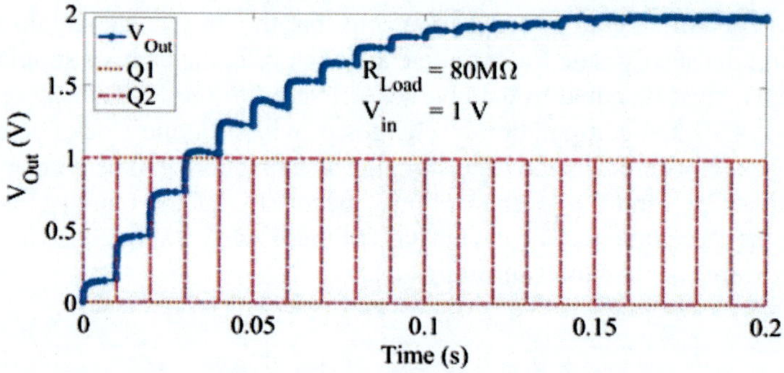

Fig. 7 Transient response of the CCVD for the fixed value of load resistance (R_L) with non-overlapping control scheme

50.76 ms, and 43.4 ms for input voltages of 1 V, 1.25 V, 1.5 V, 1.75 V, and 2 V, respectively. Moreover, the peak output voltage without load is 3.89 V, which is 97.25% of the ideal output voltage. Table 2 shows the values of the output voltage of CCVD for five different values of input voltage, such as 1 V, 1.25 V, 1.5 V, 1.75 V, and 2.0 V for the case of MOSFET, FinFET, and our work, i.e., tunnel-FET.

Further, Fig. 7 shows the characteristics of the output voltage of the proposed CCVD design concerning time for the fixed value of R_L along with non-overlapping clock signals. It shows that the process of input voltage modulated with clock signals and capacitors are used to double the input voltage after the minimum clock pulse count. Figure 8 shows the efficiency of the proposed design under different R_L for various values of drain voltage. As the design is dedicated to low-power applications, the load sweeps from 60 to 80 M Ω.

The proposed design performs better than other designs when the R_L is approximately 75 MΩ. In this regard, a comparison of output voltage concerning V_{in} is analyzed for the proposed design based on TFET and MOSFET, as shown in Fig. 9. For a fair comparison of both the designs, we have used the same design parameters, such as flying capacitors, output capacitors, and the non-overlap clock signals. Results indicate that the CCVD based on proposed Cyl GAA-TFET performs well as compared to the MOSFET-based design. Based on the above discussions, the

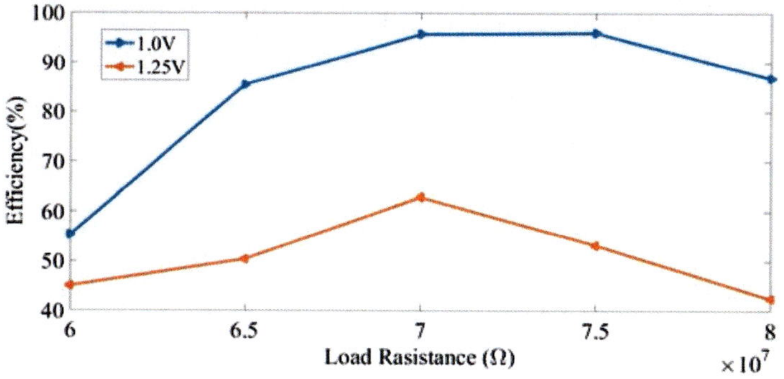

Fig. 8 The efficiency of the proposed design for varying RL under different values of drain voltage

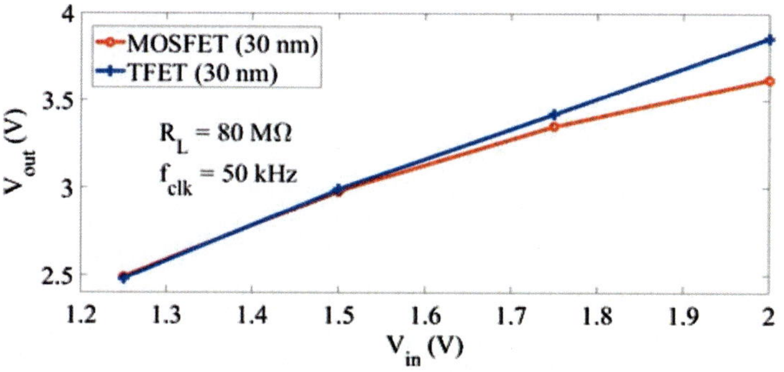

Fig. 9 Performance parameters of the CCVD design, the comparison of transfer characteristics of output voltages for the proposed design based on TFET and MOSFET

designed voltage doubler would be more beneficial for the IoT system deployed at remote locations.

5 Conclusion

In this paper, an energy-efficient CCVD based on asymmetry of underlap (AU) Cyl GAA-TFET with improved reliability has been investigated systematically in terms of device circuit co-design performance for IoT perspective. The presented design nullifies the reversion leakage current and improves high driving capability using two non-overlapping clock signals on the transistors based on the BTBT mechanism with optimized device physics. In short, it would be concluded that the proposed design based on low yield power is energy efficient, simple in fabrication, cost-effective, and implemented on a 25 nm technology node. The simulation results and comparison conclude that optimum power efficiency of the proposed design model is

95.4%. Moreover, circuit not needed extra components such as power FETs, buffers, or extra capacitor. It produces low leakage current, high driving current, and life performances compared to the latest report. Thus, in the future, the proposed design would be beneficial in a multiple-stage implementation for the system's low yield power module on chip applications due to low yield power and a small area with improved efficiency at the same time.

Acknowledgements This work was supported by the VIT Bhopal University and IIT Indore under sponsored research project of Scientific Pool Scheme of CSIR, HRDG, New Delhi.

References

1. Li, X., Ma, K., George, S., Sampson, J., Narayanan, V.: Enabling internet-of- things: opportunities brought by emerging devices, circuits, and architectures. In: 2016 IFIP/IEEE International Conference on Very Large Scale Integration (VLSI-SoC), pp. 1–6. IEEE (2016)
2. Buccolini, L., Conti, M.: An energy harvester interface for self-powered wireless speed sensor. IEEE Sens. J. **17**(4), 1097–1104 (2017)
3. Huang, Q., Jia, R., Chen, C., Zhu, H., Guo, L., Wang, J., Wang, J., Wu, C., Wang, R., Bu, W., et al.: First foundry platform of complementary tunnel-FETs in CMOS base-line technology for ultralow-power IoT applications: manufacturability, variability and technology roadmap. In: 2015 IEEE International Electron Devices Meeting (IEDM), pp. 22–26. IEEE (2015)
4. Blaauw, D., Sylvester, D., Dutta, et al.: IoT design space challenges: circuits and systems. In: 2014 Symposium on VLSI Technology: Digest of Technical Papers, pp. 1–2. IEEE (2014)
5. Liu, H., Li, X., Vaddi, R., Ma, K., Datta, S., Narayanan, V.: Tunnel FET RF rectifier design for energy harvesting applications. IEEE J. Emerg. Sel. Top. Circuits Syst. **4**(4), 400–411 (2014)
6. Beohar, A., Shah, A.P., Yadav, N.K., Vishvakarma, S.K.: Analog/RF characteristics of a 3D-cyl underlap GAA-TFET based on a Ge source using fringing-field engineering for low-power applications. J. Comput. Electron. **17**(4), 1650–1657 (2018)
7. Ionescu, A.M., Riel, H.: Tunnel field-effect transistors as energy-efficient electronic switches. Nature **479**(7373), 329–337 (2011)
8. Rewari, S., Nath, V., Haldar, S., Deswal, S., Gupta, R.S.: A novel design to improve band to band tunneling and to reduce gate induced drain leakages (GIDL) in cylindrical gate all around (GAA) MOSFET. Microsyst. Technol. (Springer) **25**, 1537–1546 (2017)
9. Avci, U.E., Morris, D.H., Young, I.A.: Tunnel field-effect transistors: prospects and challenges. IEEE J. Electron. Devices Soc. **3**(3), 88–95 (2015)
10. Beohar, A., Vishvakarma, S.K.: Performance enhancement of asymmetrical underlap 3d-cylindrical GAA-TFET with low spacer width. Micro Nano Lett. **11**(8), 443–445 (2016)
11. Kim, S.H., Agarwal, S., Jacobson, Z.A., Matheu, P., Hu, C., Liu, T.-J.K.: Tunnel field effect transistor with raised germanium source. IEEE Electron. Device Lett. **31**(10), 1107–1109 (2010)
12. Mui, T.W., Ho, M., Mak, K.H., Guo, J., Chen, H., Leung, K.N.: An area-efficient 96.5%-peak-efficiency cross-coupled voltage doubler with minimum supply of 0.8 V. IEEE Trans. Circuits Syst. II Express Briefs **61**(9), 656–660 (2014)
13. Yu, X., Moez, K., Wey, I.-C., Sawan, M., Chen, J.: A fully-integrated multistage cross-coupled voltage multiplier with no reversion power loss in standard CMOS process. IEEE Trans. Circuits Syst. II: Express Briefs (2017)
14. Kim, J.-Y., Jun, Y.-H., Kong, B.-S.: CMOS charge pump with transfer blocking technique for no reversion loss and relaxed clock timing restriction. IEEE Trans. Circuits Syst. II Express Briefs **56**(1), 11–15 (2009)

15. Vandooren, A., Leonelli, D., Rooyackers, R., Arstila, K., Groeseneken, G., Huyghebaert, C.: Impact of process and geometrical parameters on the electrical characteristics of vertical nanowire silicon n-TFETs. Solid-State Electron. **72**, 82–87 (2012)
16. Moselund, K., Ghoneim, H., Bjork, M., Schmid, H., Karg, S., Lortscher, E., Riess, W., Riel, H.: Comparison of VLS grown Si NW tunnel FETs with different gate stacks. In: Proceedings of the European Solid State Device Research Conference, 2009. ESSDERC'09, pp. 448–451. IEEE (2009)
17. Beohar, A., Yadav, N.K., Vishvakarma, S.K.: Analysis of trap assisted tunneling in asymmetrical underlap 3D-cylindrical GAA-TFET based on hetero-spacer engineering for improved device reliability. IET Micro Nano Lett. **12**(12), 982–986 (2017)

Impact of Channel Epilayer Induced Corner-Effect on the Sensing Performance of a Unique PTFET-Based Biosensor (Epi-PTFET-Biosensor)

Sanu Gayen, Suchismita Tewari, and Avik Chattopadhyay

1 Introduction

Nowadays, bio-sensing, or precisely the in-time detection of various bio-molecules, has gained immense importance as well as utmost urgency in day-to-day human life. This sudden surge in bio-sensing demand in various fields, viz., medical diagnosis, environment monitoring, etc., coupled with the requirement of sensors, having the qualities, like quick response time, user-friendliness, cost-effectiveness, and simple enough for mass production, have perked the interest in the field of interdisciplinary research, termed as bio-sensing-technology to a great degree [1–3]. The economic label-free detection along with compatibility with traditional CMOS process-flow and easy scalability have made the FET-based biosensors a lucrative choice for the said purpose.

The device miniaturization has been continued since the last few decades in order to increase package density while keeping the device footprint small enough to match the insane market demand of easy portability interlaced with multi-functionality in a single system. This leads to the incorporation of the increased number of devices in a single system, which urges the individual device to consume less and less power and to work under extremely scaled-down supply voltage and thus, ultimately become suitable for low-power electronics. Tunnel FET (TFET) devices are the fine match to fulfill all the above-mentioned criteria having sub-60 mV/decade subthreshold swing (SS) at room temperature [4, 5].

A number of works have been reported earlier, using advanced device architectures, alternative channel materials, etc., in order to enhance the sensitivity of the TFET-biosensors. However, all of those schemes lead to high cost and high fabrication complexity [6–8]. While incorporating different architectural schemes, 'L'-shaped and 'U'-shaped gated structures have gained lots of popularity, allowing to use the vertical tunneling feature, hence enhancing the device ON-current. But, the

S. Gayen (✉) · S. Tewari · A. Chattopadhyay
Department of Radio Physics and Electronics, University of Calcutta, Kolkata, India

A. Biswas et al. (eds.), *Microelectronics, Circuits and Systems*, Lecture Notes in Electrical Engineering 755, https://doi.org/10.1007/978-981-16-1570-2_8

well-known corner-effect (a crowding of electric field lines, both in horizontal as well as the vertical direction, at the corner point, resulting in a sudden upsurge of electric field strength at corner region, which gives rise to a certain spike of current and mainly affects the subthreshold region) imposed by these unique structures and their impact on device characteristics have seldom analyzed in detail.

In this paper, for the first time, we have introduced an electronic pTFET-based biosensor, with an inverted 'L'-shaped ('⅂'-shaped) channel, by introducing a channel epilayer beneath the nanogap cavity, and have named the device as epi-pTFET-biosensor. A detailed investigation has been carried out in order to study the impact of this epilayer, hence the generated corner-effect by this unique shape, on the sensitivity and detectability performance of the proposed epi-pTFET-biosensor device, followed by determining the end-point of the nanogap cavity and optimum nanogap-cavity length for a wide variety of bio-molecules, viz., Protein-G, Ferrocytochrome-C, Ferricytochrome-C, Myoglobin, and Apomyoglobin. The channel material is restricted to Silicon to keep the device economic. The device architecture is also a minor diversion from the standard bulk or SOI architecture, without adding up fabrication complexity. Interestingly, we have found that this subtle, yet effective, architectural modification has ultimately paid off by turning the undesired corner-effect into a favorable one by giving improved detectability with satisfactory sensitivity performance for the proposed epi-pTFET-biosensor device.

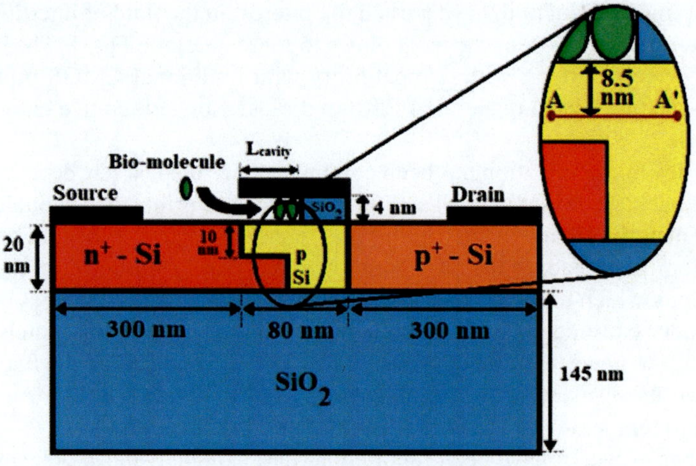

Fig. 1 2D schematic of the proposed epi-pTFET-biosensor device

2 Device Structure and Simulation Scheme

The proposed epi-pTFET-biosensor schematic is shown in Fig. 1. The gate oxide thickness of the proposed device is kept as 4 nm, for the incorporation of one monolayer of Myoglobin or Apomyoglobin, having the highest diameter among all the aforementioned bio-molecules. A channel epilayer of thickness 10 nm is considered in our study. A portion of gate oxide is etched (starting from the left most source/channel junction (at the surface) to slightly beyond the corner of the 'T'-shaped channel), to make a nanogap cavity. For gate metal, TaN is chosen containing a work function of 4.15 eV [9]. The doping density of the n-type source and p-type drain region are considered as $10^{22}/cm^3$ and 10^{19} cm^{-3}, respectively.

A numerical device simulator, SILVACO ATLAS [10], is used to simulate the proposed epi-pTFET-biosensor. The model, named, Non-local BTBT (band-to-band tunneling) is incorporated in the simulation to capture the spatial change within the energy bands. The Non-local BTBT model, as applicable to our aforesaid numerical device simulator [10], incorporates a sophisticated tunneling transmission (T_{WKB}) following WKB (Wentzel–Kramers–Brillouin) approximation [11], computed accurately based on the actual potential profile variation throughout the tunneling barrier rather than by assuming constant or average electric field along that path. The complex band structure inside the forbidden energy gap, as well as the ellipticity, along with the dual electron/hole behavior of tunneling carriers can be absorbed in m_t by setting it almost equal to $0.7m_r$ [11], where m_t and m_r are the tunneling effective mass and reduced effective mass, respectively, which ultimately leads to the tunneling transmission as [11]

$$T_{WKB} = \exp\left[-\frac{\pi}{qF}\frac{\sqrt{(0.7m_r)}}{2\hbar}E_g^{3/2}\right] \tag{1}$$

where $\hbar$ stands for the reduced Planck constant, q stands for the elementary charge, F stands for the electric field in the transition region from tunnel-source to tunnel-destination, E_g stands for the bandgap of the semiconductor, $m_t = (0.7\ m_r)$ is the "tunneling effective mass", and m_r is the "reduced effective mass". Additionally, Auger and SRH (Shockley–Read–Hall) models are invoked to capture the effect of carrier recombination and generation. Bandgap narrowing model is integrated to incorporate the high concentration of doping and finally, the field (lateral as well as longitudinal)-, concentration-, and temperature- dependent mobility is modeled by invoking CVT Lombardi mobility model. FD (Fermi–Dirac) statistics is used all through the simulation.

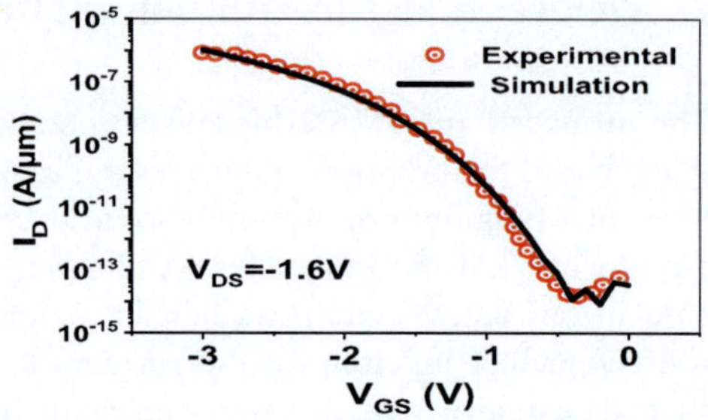

Fig. 2 Transfer characteristics comparison between the simulated pTFET device and the experimentally fabricated device [12]

3 Model Calibration

At first, the validation of our simulation scheme has been done against the characteristic of an experimentally fabricated pTFET device of Ref. [12]. The simulated device architecture, dimensions, as well as materials are chosen exactly the same as in the case for the reported fabricated pTFET [12]. Figure 2 demonstrates a comparison between the transfer characteristics of the device in Ref. [12] and the simulated pTFET device. Clearly, the simulated characteristics tally with that of the experimental characteristics of the device in Ref. [12] in an excellent manner, as is evident from Fig. 2, confirming the genuineness of the basis of our simulation.

4 Results and Discussion

The successful recognition of bio-molecule(s) through some equivalent identified electronic or material parameters, using which an alteration or deviation of device characteristics takes place, indicating the presence of the bio-molecule(s), is the essence of electronic bio-sensing mechanism. In this study, the dielectric constant (k) of bio-molecule(s) is identified as the equivalent material parameter, and throughout the study, each type of bio-molecules, with its specified diameter and k-value [13], is modeled as an insulator. This section is subdivided into two parts. In the first part, the end-point of the nanogap cavity is determined, by exploring the electric field profile along the channel (considering the corner-effect) of the proposed epi-pTFET-biosensor device. In the second part, a rigorous analysis of sensitivity parameters of the proposed device has been studied, followed by determining the optimum length-window of nanogap cavity ($OLW_{(ncavity)}$) for five different types of bio-molecules, and detectability (D) [i.e., minimum number of bio-molecules required for detecting its presence], for all the five aforementioned bio-molecules. Table 1 has tabulated the information regarding the above-mentioned bio-molecule diameters and corresponding dielectric constant values [13].

Table 1 Different parameters of the bio-molecules [13]

Types of bio-molecules	K-value	Diameter (nm)
Protein-G	4.0	2.66
Ferrocytochrome-C	24.0	3.30
Ferricytochrome-C	3.4	3.18
Myoglobin	3.5	4.00
Apomyoglobin	6.0	4.00

4.1 Determination of the End-Point of the Nanogap Cavity

Figure 3a shows the device schematic without the nanogap cavity. Figure 3b shows the electric field profile along the horizontal cutline (BB′), shown in Fig. 3a. It is observed that there is a sudden rise of electric filed, at the mid-portion of the channel.

This may be attributed to the fact of the crowding of electric field lines due to the corner-effect, at the sharp down-right corner of the ‘Ͳ’-shaped channel. This sharp rising of the electric field, in the mid-portion of channel (~5 nm on both sides of the corner point, with a peak value at the corner point) makes that region highly sensitive against any kind of external stimulant. It is expected that any kind of modification of the k-value of gate oxide will affect that sensitive zone a lot more than that for the rest of the channel region. This makes the aforementioned region most appropriate for the sensing purpose. Based on this concept, the right end-point of the nanogap cavity, required for capturing the bio-molecules, has been chosen at the right side of that corner point by a distance equal to twice the highest diameter of the bio-molecules, covering the entire sensitive zone, as is shown in Fig. 1, so that the incorporation of a few numbers of bio-molecules in the cavity above that sensitive zone gives rise to an appreciable change in the targeted output parameters, accelerating the detection procedure and hence helping to improve the corresponding detectability.

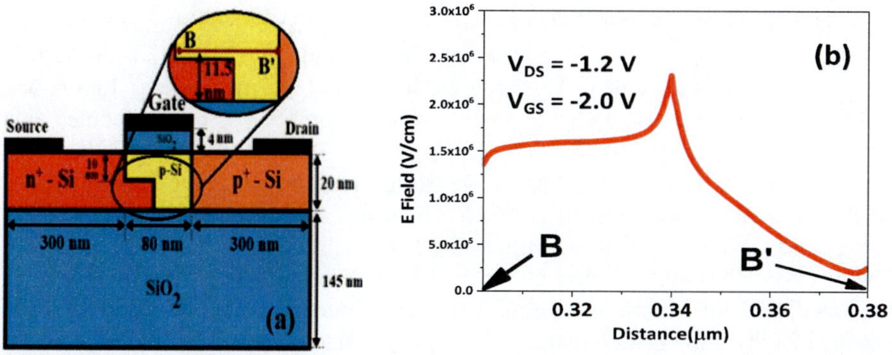

Fig. 3 **a** Device schematic without the nanogap cavity, **b** corresponding electric field profile along BB′

4.2 Detectability and Sensitivity Study of the Proposed Epi-PTFET-Biosensor Device

In this exploration, the slope-change and shift of the device characteristics, due to the existence of the bio-molecule(s) in the nanogap cavity have been taken into account, and based on those phenomena, the following five sensitivity parameters have been chosen for further performance analysis, viz.,

(a) Threshold voltage shift

$$[\Delta \mathrm{V_{th}}] = V_{th(Empty\,cavity)} - V_{th(bio-molecules)} \quad (2)$$

(b) Threshold voltage sensitivity

$$[\mathrm{S(V_{th})}] = \frac{V_{th(Empty\,cavity)} - V_{th(bio-molecules)}}{V_{th(Empty\,cavity)}} \times 100\% \quad (3)$$

(c) Drive current shift

$$[\Delta \mathrm{I_{DS}}] = I_{DS(Empty\,cavity)} - I_{DS(bio-molecules)} \quad (4)$$

(d) Drive current sensitivity

$$[\mathrm{S(I_{DS})}] = \frac{\left[log_{10} I_{DS(Empty\,cavity)} - log_{10} I_{DS(bio-molecules)}\right]}{log_{10} I_{DS(Empty\,cavity)}} \times 100 \quad (5)$$

(e) Subthreshold swing sensitivity

$$[\mathrm{S(SS)}] = \frac{SS_{(Empty\,cavity)} - SS_{(bio-molecules)}}{SS_{(Empty\,cavity)}} \times 100 \quad (6)$$

In the parameters, defined in (c) and (d), $I_{DS(Empty\ cavity)}$ denotes the current at $V_{GS} = \{V_{th(Empty\ cavity)} + (0.67) \times V_{DD}\}$ for empty cavity and $I_{DS(bio\text{-}molecules)}$ is the current at the aforementioned value of V_{GS} after the addition of bio-molecule(s). Figure 4a–e shows the set of transfer characteristics with the number of bio-molecules as a parameter, for each type of bio-molecules. In each case, it is observed that with the addition of bio-molecules, one by one from the right end-point of the nanogap cavity, the transfer characteristic shifts slowly toward the right, hence decreasing the corresponding device threshold voltage or a right-shift of V_{th} has been observed. The degree of this shifting decreases gradually as the number of bio-molecules is increased, and ultimately, after adding a certain number of bio-molecules in the nanogap cavity, the characteristics are seen to overlap with its immediate previous one, and further incorporation of bio-molecules does not shift the characteristic anymore. This is termed as shift-saturation case. This may be attributed to the fact that, as long as the position of the bio-molecules in the nanogap cavity falls right

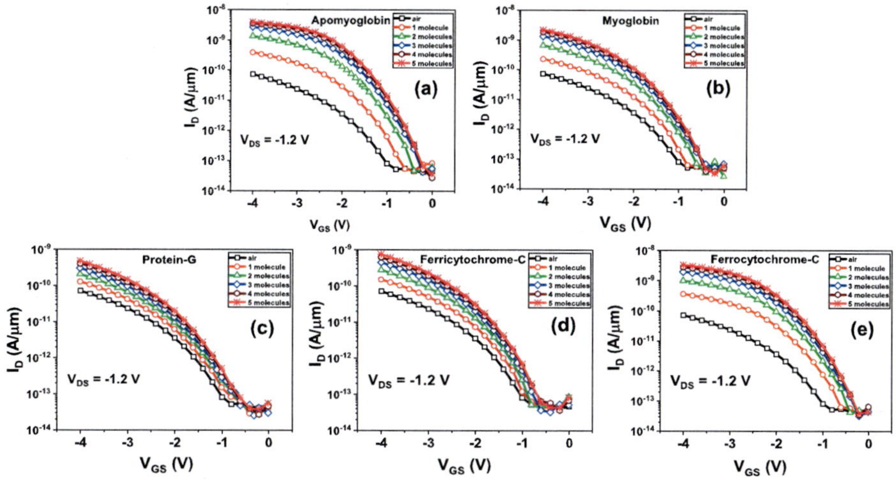

Fig. 4 The epi-pTFET-biosensor device—characteristics (transfer) of the conjugated bio-molecules [**a** Apomyoglobin, **b** Myoglobin, **c** Protein-G, **d** Ferricytochrome-C, **e** Ferrocytochrome-C] and of empty cavity with the number of the bio-molecules as a parameter

above the previously discussed sensitive zone or corner-field zone of the channel, the presence of even just one molecule results in a sharp increase in the corner-field resulting an upsurge in hole tunneling (combined effect of vertical and horizontal tunneling) rate from source to channel in that corner region. This shifts the entire characteristics toward the right, decreasing the corresponding V_{th}. With the addition of each bio-molecule, the position of the new one in the nanogap cavity is shifting away from the right end-point of the cavity, moving through the sensitive zone, and ultimately crossing the region. In each step of such bio-molecule addition, the impact of the new molecule on the device characteristic gradually decreases, causing a gradual decrement of right-shift of the characteristics, and ultimately achieving shift-saturation condition. This phenomenon is reflected from Fig. 5a, b, showing the corresponding electric field profiles, along with the hole tunneling rate for Apomyoglobin and Ferrocytochrome-C (along the horizontal line AA′ in Fig. 1). A similar trend has been followed in the case of the rest of the bio-molecules. The five aforementioned sensitivity parameters are extracted for each type of bio-molecules and are tabulated in Table 2. Based on that, the minimum cavity length ($LC_{(min)}$) in each case is determined. Interestingly, it has been found that for each type of bio-molecules, only one number of the corresponding molecule is enough to give a measurable value of all the five sensitivity parameters. This leads to enhanced detectability of the bio-molecules by the proposed epi-pTFET-biosensor device, requiring only one molecule of each type of bio-molecules for successful detection. It is to be further noted that, in each case, this initial shift, for each type of bio-molecules, well surpasses the corresponding change (in value) due to input (gate) voltage fluctuation (around $V_{GS} = V_{th}$) by a couple of order of thermal voltage, and thus preventing any sort of false detection by minor thermal fluctuation. This ensures the reliability and hence stability of

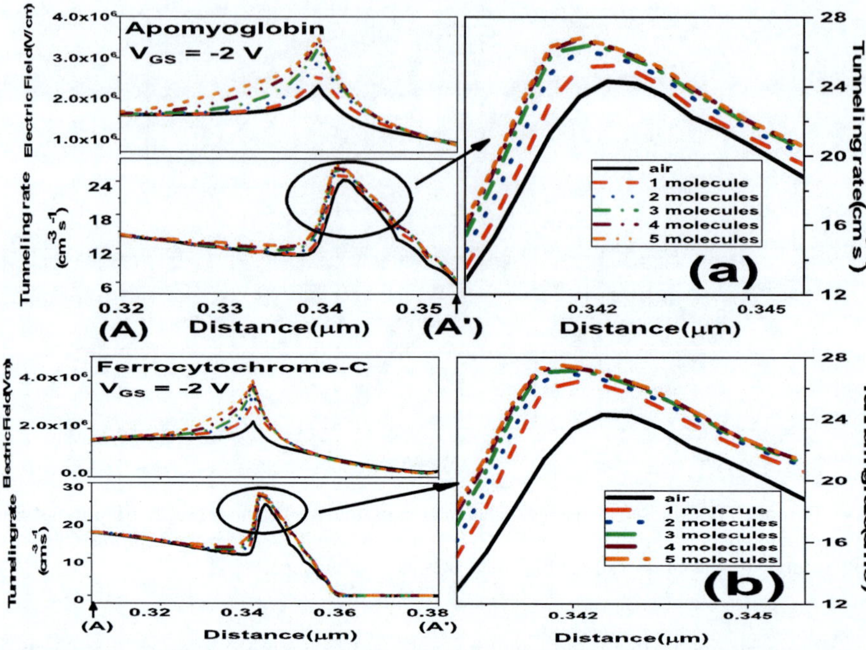

Fig. 5 Electric field profiles and tunneling rates of epi-pTFET-biosensor device along AA′ for empty cavity and for [**a** Apomyoglobin, **b** Ferrocytochrome-C] with the number of the bio-molecules as a parameter

the proposed epi-pTFET-biosensor device against thermal noise. Hence, to cover the detection of the entire range of five types of bio-molecules, the minimum nanogap-cavity length should be equal to [D × (diameter of the largest bio-molecule)], which happens to be 4 nm. With the addition of further bio-molecules of each type, all five sensitivity parameters are observed to be increased, although the degree of increment decreases with each addition, and ultimately it becomes non-significant or what we have termed as achieving shift-saturation condition. It is observed that for each type, adding beyond five molecules, does not give any significant change, especially in the case of ΔV_{th} and $S(V_{th})$ parameters. Thus, the five number of molecule, of each type, is considered as the attainment of shift-saturation and the corresponding cavity length (5 × diameter of bio-molecule) is considered to be the maximum limit of the nanogap cavity and is termed as shift-saturation cavity length ($LC_{(shift\text{-}sat)}$). Thus, an optimum length-window of the nanogap cavity ($OLW_{(ncavity)}$), considering the detectability factor as well as the shift-saturation case to incorporate increased sensitivity values, can be defined for the error-free and successful detection of the wide variety of the aforementioned bio-molecules. This parameter is tabulated in Table 2 as well. The minimum limit of $OLW_{(ncavity)}$ is the highest $LC_{(min)}$ and the maximum limit is the highest $LC_{(shift\text{-}sat)}$ of all the five types of bio-molecules, respectively. Table 3 shows a comparison between the detectability, offered by the

Table 2 Sensitivity parameters, detectability (D), and optimum length-window of nanogap cavity ($OLW_{(ncavity)}$) for five different types of bio-molecules

Types of bio-molecules	Number of bio-molecules	ΔV_{th} (V)	$S(V_{th})$ (%)	ΔI_{DS}(decade)	$S(I_{DS})$ (%)	S(SS) (%)	D	$LC_{(min)}$ (nm)	$LC_{(shift\text{-}sat)}$ (nm)	$OLW_{(ncavity)}$(nm)
Apomyoglobin	1	0.82	33.2	0.82	7.8	44.8	1	4.00	20.0	From 4 nm to 20 nm
	2	1.22	49.4	1.41	13.5	57.5				
	3	1.41	57.1	1.74	16.6	61.1				
	4	1.50	60.7	1.89	18.1	63.0				
	5	1.53	61.9	1.95	18.6	64.1				
Myoglobin	1	0.55	22.3	0.53	5.1	34.0	1	4.00	20.0	
	2	0.88	35.6	1.00	9.6	47.2				
	3	1.06	42.9	1.30	12.4	53.2				
	4	1.14	46.2	1.46	13.9	55.4				
	5	1.18	47.8	1.53	14.6	57.0				
Protein-G	1	0.28	11.3	0.25	2.4	17.8	1	2.66	13.3	
	2	0.45	18.2	0.45	4.3	27.5				
	3	0.58	23.5	0.61	5.8	35.2				
	4	0.67	27.1	0.73	7.0	39.3				
	5	0.71	28.7	0.81	7.7	40.3				
Ferri-cytochrome-C	1	0.35	14.2	0.32	3.1	22.9	1	3.18	15.9	
	2	0.57	23.1	0.58	5.5	35.2				
	3	0.72	29.2	0.79	7.5	42.2				
	4	0.80	32.4	0.92	8.8	44.8				
	5	0.85	34.4	1.00	9.6	46.4				

(continued)

Table 2 (continued)

Types of bio-molecules	Number of bio-molecules	ΔV_{th} (V)	$S(V_{th})$ (%)	ΔI_{DS}(decade)	$S(I_{DS})$ (%)	S(SS) (%)	D	$LC_{(min)}$ (nm)	$LC_{(shift\text{-}sat)}$ (nm)	$OLW_{(ncavity)}$(nm)
Ferro-cytochrome-C	1	0.85	34.4	0.82	7.8	44.8	1	3.30	16.5	
	2	1.13	45.8	1.29	12.3	53.2				
	3	1.29	52.2	1.59	15.2	58.9				
	4	1.36	55.1	1.75	16.7	60.2				
	5	1.40	56.7	1.83	17.5	61.9				

Table 3 Comparison of detectability (D) between the biosensor device of this work and that of the work in [14]

Name of the bio-molecules	Detectability (D)	
	Proposed epi-pTFET-biosensor	Traditional SOI pTFET sensor device [14]
Apomyoglobin	1	2
Myoglobin	1	2
Protein-G	1	3
Ferricytochrome-C	1	3
Ferrocytochrome-C	1	3

proposed epi-pTFET-biosensor device, as opposed to a more traditional, purely SOI pTFET sensor device [14]. From the comparison, it is clearly seen that our proposed epi-pTFET-biosensor device outperforms the traditional sensor device by a great degree, for each type of bio-molecules, claiming its absolute superiority from the detectability point-of-view. This leads to the reduction of the minimum length of nanogap cavity by two times in the case of Myoglobin and Apomyoglobin and three times in case of Ferrocytochrome-C, Ferricytochrome-C, and Protein-G, respectively, for the proposed epi-pTFET-biosensor device over the equivalent conventional SOI pTFET sensor device, leading to reduced device footprint and ultimately, establishing the triumph of the introduced subtle, yet effective, architectural modification of the proposed device.

5 Conclusion

In this paper, a unique epi-pTFET-biosensor device has been proposed and a rigorous analysis, regarding the performance of the aforementioned device in the sensory domain, has been carried out in terms of five different sensitivity parameters, and based on that analysis, the detectability and an optimum length-window of nanogap cavity for the detection of all the discussed bio-molecules have been determined. Interestingly, it has been observed that the corner-effect, usually proved to be undesired, here, actually comes to the aid of the proposed sensor device in enhancing its detectability to a great degree, compared to its equivalent traditional SOI pTFET sensor device, for a wide range of bio-molecules.

References

1. Luong, J.H.T., Male, K.B., Glennon, J.D.: Biosensor technology: technology push versus market pull. Biotechnol. Adv. **26**(5), 492–500 (2008)
2. Kim, S., Baek, D., Kim, J.-Y., Choi, S.-J., Seol, M.-L., Choi, Y.-K.: A transistor-based biosensor for the extraction of physical properties from biomolecules. Appl. Phys. Lett. **101** 073703-1-4 (2012)
3. Barbaro, M., Bonfiglio, A., Raffo, L.: A charge-modulated FET for detection of biomolecular processes: conception, modeling, and simulation. IEEE Trans. Electron Devices **53**(1), 158–166 (2006)
4. Ribaut, C., Loyez, M., Larrieu, J.C., Chevineau, S., Lambert, P., Remmelink, M., Wattiez, R., Caucheteur, C.: Cancer biomarker sensing using packaged plasmonic optical fiber gratings: Towards in vivo diagnosis. Biosens. Bioelectron **92**, 449–456 (2017)
5. Shin, J., Choi, S., Yang, J.-S., Song, J., Choi, J.-S., Jung, H.-I.: Smart forensic phone: colorimetric analysis of a bloodstain for age estimation using a smartphone. Sens. Actuators B, Chem. **243**, 221–225 (2017)
6. Kanungo, S., Chattopadhyay, S., Gupta, P.S., Rahaman, H.: Comparative performance analysis of the dielectrically modulated full-gate and short-gate tunnel FET-based biosensors. IEEE Trans. Electron Devices **62**(3), 994–1001 (2015)
7. Kalra, S., Jagadesh Kumar, M., Dhawan, A.: Dielectric-modulated field effect transistors for DNA detection: impact of DNA orientation. IEEE Electron Device Lett. **37**(11), 1485–1488 (2016)
8. Narang, R., Reddy, K.V.S, Saxena, M., Gupta, R.S., Gupta, M.: A dielectric-modulated tunnel-FET-based biosensor for label-free detection: analytical modeling study and sensitivity analysis. IEEE Trans. Electron Devices **59**(10), 2809–2817 (2012)
9. Lee, B.H., Choi, R., Kang, L., Gopalan, S., Nieh, R., Onishi, K., Jeon, Y., Qi, W.-J., Kang, C., Lee, J.C.: Characteristics of TaN gate MOSFET with ultra thin hafnium oxide (8Å–12Å). IEDM Tech. Dig. 39–42 (2000)
10. ATLAS User's Manual: A Device Simulation Package. SILVACO Int, Santa Clara, CA (2015)
11. Ilatikhameneh, H., Salazar, R., Klimeck, G., Rahman, R., Appenzeller, J.: From Fowler–Nordheim to nonequilibrium Green's function modeling of tunneling. IEEE Trans. Electron Devices **63**(7), 2871–2878 (2016)
12. Migita, S., Fukuda, K., Morita, Y., Ota, H.: Experimental demonstration of temperature stability of Si-tunnel FET over Si-MOSFET. In: IEEE Silicon Nanoelectronics Workshop (SNW) (2012)
13. Simson, T., Brooks, C.: Charge screening and dielectric constant of proteins: insights from Molecular Dynamics. J. Am. ChemSoc. 8452–8458 (1996)
14. Gayen, S., Tewari, S., Chattopadhyay, A.: Exploring p-channel TFET for optimum cavity-length window in detecting a wide variety of protein-molecules with the effect of their position dependent variability on sensitivity. IETE Tech. Rev. 1–10 (2020). https://doi.org/10.1080/02564602.2020.1747951

Performance Comparison of III–V and Silicon FinFETs for Ultra-Low Power VLSI Applications

Ankit Dixit, Dip Prakash Samajdar, Vibhuti Chauhan, and Navjeet Bagga

1 Introduction

Since the last few decades, miniaturization of the device technology is one of the biggest challenges for the researcher to uphold Moore's law of scaling [1] and thus to continue follow this era of 'More than Moore' [2, 3]. Numerous novel device physics and geometrical modifications in the device have been timely proposed. Tunnel Field-Effect Transistor (TFET) [4], Impact Ionization MOS (IMOS) [5, 6], and Junctionless FET [7] are some innovative efforts in this domain which were proposed and carried out by researchers. Planar silicon process technology also reached its saturation and therefore the inclusion of new device material has been investigated and proposed as switching transistors [8, 9]. Among new channel material, III–V compound semiconductor (SC) has great potential because of having high mobility and the property to tune the bandgap by changing the mole fraction of the compounds [10–12]. In the available literature, many experimental and simulation results of InGaAs-FinFET [13, 14] and InAs [15, 16] have been discussed solely which claims the potential of the promising candidacy for III–V SC in future FinFET technology. However, a suitable comparison of employing III-V SC over conventional silicon as a channel material is still missing. In this paper, DC and RF performance is investigated for both group IV (Silicon) and III–V (InAs and $In_xGa_{1-x}As$) as a channel material in FinFET. The paper is organized as follows: in Sect. 2, the device structure and simulation framework are discussed. Section 3 discusses the acquired results of our study and finally, Sect. 4 summarizes the paper.

A. Dixit · D. P. Samajdar (✉) · V. Chauhan · N. Bagga
VLSI Design and Nano Scale Computational Lab, PDPM Indian Institute of Information Technology Design and Manufacturing, Jabalpur 482005, India
e-mail: dipprakash010@gmail.com

N. Bagga
e-mail: dipprakash010@gmail.com

A. Biswas et al. (eds.), *Microelectronics, Circuits and Systems*, Lecture Notes in Electrical Engineering 755, https://doi.org/10.1007/978-981-16-1570-2_9

2 Device Structure and Simulation Framework

Figure 1 shows a three-dimensional (3D) view of FinFET of silicon-on-insulator (SOI) configuration adopted for this study. In this work, a commercially available Sentaraus TCAD tool is used for numerical simulations [17, 18]. All the simulation has been carried out for gate length (LG) 22nm, Fin height (HFIN) 30nm, and Fin Width (WFIN) of 10nm. In Tri-Gate devices like FinFFET, the drain current is directly proportional to the effective device width which is the sum of twice of fin height and fin width. Titanium Nitrate (TiN) is used as a gate contact due to its low resistivity and silicon nitrate ($Si3N_4$) is used as a spacer between the source/drain (S/D) pad and the gate to reduce the strain effect in the device. High-k material (HfO_2) with dielectric constant (Ɛ) of 22 has been considered as a gate dielectric to increase the gate controllability and reduce the probability of gate tunneling current.

Device simulations have been performed using Low-Doped Drain (LDD) architecture to reduce the effect of a high electric field near the junction [19]. Table 1 enlists the material properties of III–V SC and silicon at 300 K [17]. The default

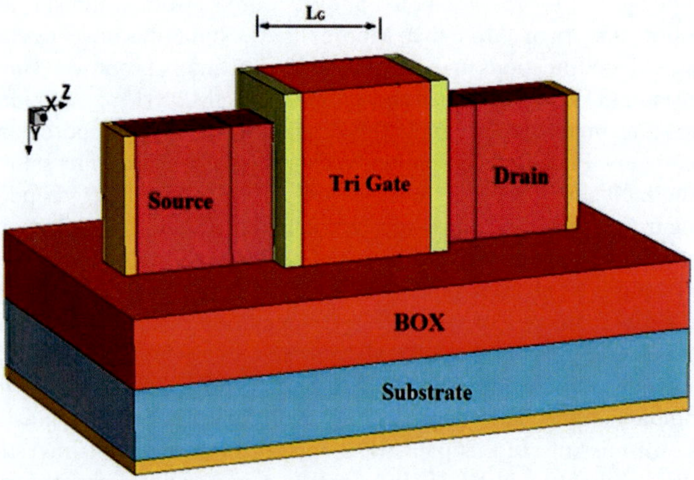

Fig. 1 A three-dimensional view of FinFET device based on SOI configuration

Table 1 Electrical parameters of different materials at 300 k

S. no	Parameter	Silicon	InAs	$In_xGa_{1-x}As$ ($x = 0.53$)
1	Ɛ	11.70	14.55	13.90
2	μ_{nmax} (cm^2V^{-1} s^{-1})	1.417×10^3	2.26×10^4	1.59×10^4
3	μ_{pmax} (cm^2V^{-1} s^{-1})	4.705×10^2	2.50×10^2	3.20×10^2
4	E_g (eV)	1.169	0.415	0.718
5	n_i (cm^{-3})	1.5×10^{10}	1×10^{15}	6.3×10^{11}

Table 2 Parameters used in simulations

S. no	Parameter	Symbol	Value
1	D/S Doping (cm^{-3})	N_D	1×10^{20}
2	Channel Doping (cm^{-3})	N_A	1×10^{16}
3	Drain Length (nm)	L_D	10
4	Source Length (nm)	L_S	10
5	Channel Length (nm)	L_G	22
6	Fin height (nm)	H_{FIN}	30
7	Fin width (nm)	W_{FIN}	10
8	Substrate Thickness (nm)	t_{SUB}	10
9	BOX Thickness (nm)	t_{BOX}	15
10	Gate Oxide Thickness (EOT)(nm)	t_{ox}	1
11	Spacer Thickness (nm)	t_{sp}	2
12	Gate Work Function (eV)	φ_G	4.52

values of device dimensional parameters used in this study are mentioned in Table 2, unless stated otherwise. In our simulation framework, we have used drift–diffusion models, doping dependence mobility models along with Augur and SRH recombination models. In addition, radiative recombination models have also been considered for III–V materials. Our simulation models are well-calibrated with the reported experimental data [20] for low and high drain to source (VDS) values as shown in Fig. 2.

3 Results and Discussion

To investigate the DC behavior of the device, transfer characteristics (ID-VGS) are plotted for distinct channel material employed in FinFET as depicted in Fig. 3. A negative threshold voltage for InAs/ InGaAs and a positive threshold voltage for silicon are observed. III–V materials are having a low bandgap which can causes a high leakage current in the device and in turn provide a low ON to OFF current (I_{ON}/I_{OFF}) ratio. Whereas, due to higher mobility, the ON current level is high as compared to conventional silicon. Therefore, III–V material-based devices suffer from poorer subthreshold slope (SS) and result in an inadequate strength to adopt these materials over silicon technology. However, due to the lower threshold voltage in III–V material-based devices, a maximum value of transconductance has been obtained at smaller gate voltage as depicted in Fig. 4 which results in a higher intrinsic gain of the transistor and is beneficial in analog applications.

In addition to the analog performances, the radio frequency (RF) analysis of the device provides a significant improvement in the performance of the device. In III–V-based FinFET, by applying a positive gate voltage, the channel gets easily depleted

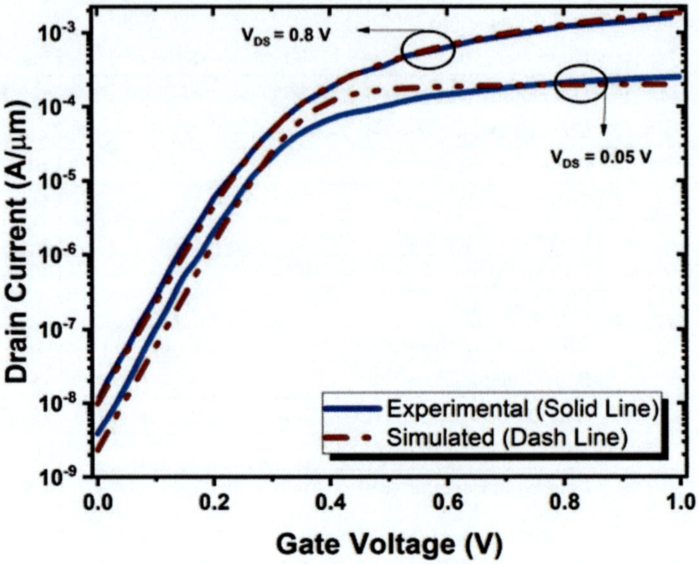

Fig. 2 Calibration of the simulation framework with reported experimental data [20] for 22 nm technology

and in turn the inversion layer beneath the channel is formed at lower voltages as compared to its silicon counterpart. The variation of the charge concentration as a function of applied gate voltage is shown in Fig. 5. It is clearly observed that for InAs higher charge concentration has been obtained due to larger intrinsic concentration in the substrate.

The overall gate capacitance (C_{GG}) is an important parameter to obtain various figures of merits (FoMs) of RF performance like static and dynamic power dissipation, energy and power delay product, intrinsic delay, and unity gain bandwidth of the device [21, 22]. Figure 6 illustrates the variation of total gate capacitance (C_{GG}) as a function of applied gate voltage. It is evident that the lower gate capacitance of III–V-based device improves the switching performance which results due to lower intrinsic delay, static power, and static energy dissipation as compared to silicon-based device. Table 3 summarizes the results of DC and RF performance parameters.

4 Conclusions

In this paper, a tri-gate FinFET is investigated using group IV and III–V material as a channel material. In comparison to III–V materials, a superior I_{ON}/I_{OFF} ratio is obtained for silicon material due to the higher energy bandgap whereas a higher value of transconductance is achieved in the case of III–V SC at a lower applied

Table 3 Comparative analysis of DC nad RF parameters for different materials used in the channel region

Performance parameters

		I_{ON}	I_{OFF}	g_m	C_{GG}	τ	SPD	SED	UGB	TGF
S. No	Material	(Amp)	(Amp)	(S)	(fF)	(ps)	(pW)	(aJ)	(GHz)	(V^{-1})
		I_D@V_G = 1.0V	I_D@V_G = 0.0V	$\frac{dI_D}{dV_G}$	C_{GG}@V_G = 1.0V	$\frac{C_{GG}V_{DD}}{I_D}$	$C_{GG}V_{DD}^2 f$	$C_{GG}V_{DD}^2$	$\frac{g_m}{2\pi C_{GG}}$	$\frac{g_m}{I_D}$
1	InAs	5.85×10^{-5}	4.09×10^{-7}	6.93×10^{-5}	6.76×10^{-2}	5.78×10^{-2}	1.69×10^{-1}	1.69×10^{-1}	1.63×10^2	1.18
2	$In_xGa_{1\text{-}x}As$ (x = 0.53)	5.33×10^{-5}	7.59×10^{-8}	1.13×10^{-4}	6.67×10^{-2}	6.26×10^{-2}	1.67×10^{-1}	1.67×10^{-1}	2.70×10^2	2.12
3	Silicon	4.58×10^{-5}	6.90×10^{-12}	1.19×10^{-4}	1.15×10^{-1}	1.26×10^{-1}	2.88×10^{-1}	2.88×10^{-1}	1.65×10^2	2.60

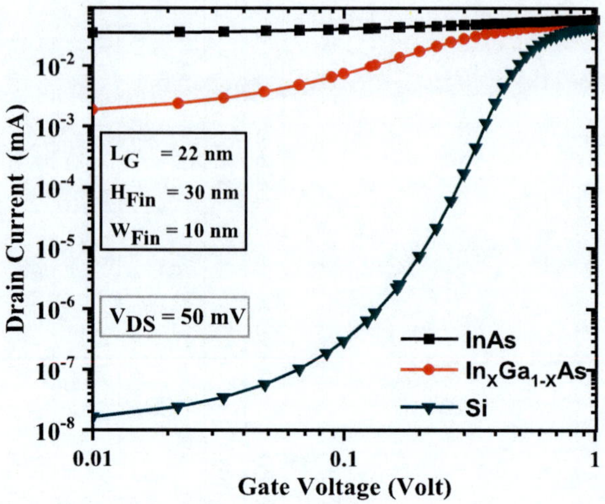

Fig. 3 Transfer characteristics (ID-VGS) of Tri-gate FinFET for distinct channel materials

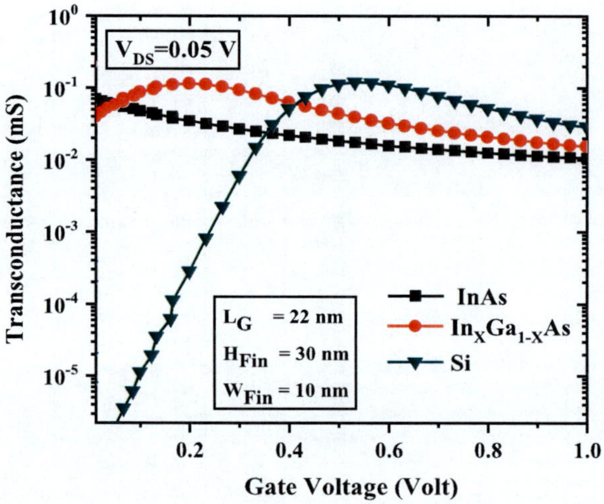

Fig. 4 Transconductance as a function of applied gate voltage in Tri-gate FinFET for distinct channel material

gate voltage. Analog performance metrics are compared for distinct materials which clearly depicts that reduction in the total gate capacitance for InGaAs and InAs materials is about 42% of its silicon counterpart. RF performance metrics such as Power dissipation, energy dissipation, and intrinsic gate delay (τ) are superior in III–V material-based devices, and a maximum value of unity gain bandwidth (UGB) of 270 GHz is achieved in the InGaAs whereas a maximum value of transconductance generation factor (TGF) of about 2.60 is obtained for the silicon-based device. The results indicate that III–V materials are having good potential for ultra-low power application whereas the DC performance of material can be further improved by advanced process technology.

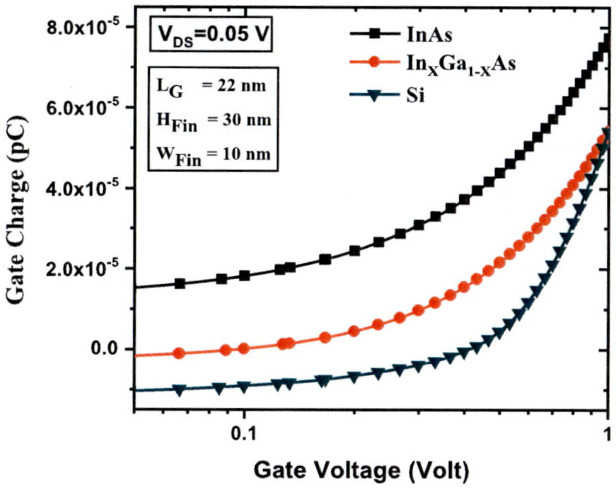

Fig. 5 Gate charge concentration as a function of applied gate voltage of Tri-gate FinFET for distinct channel material

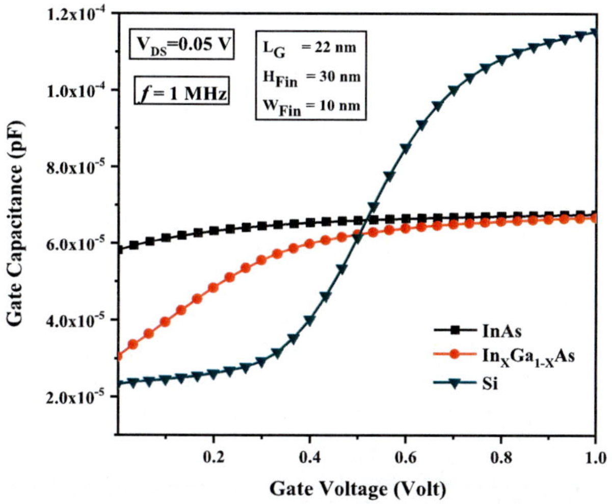

Fig. 6 Total Gate Capacitance (C_{GG}) as a function of the applied gate voltage of Tri-gate FinFET for distinct channel material

References

1. Thompson, S.E., Parthasarathy, S.: Moore's law: the future of Si microelectronics. Mater. Today **9**(6), 20–25 (2006). https://doi.org/10.1016/S1369-7021(06)71539-5
2. Waldrop, M.M.: More than Moore. Nature **530**(7589), 144–147 (2016). doi: https://doi.org/10.1038/530144a.
3. International Technology Roadmap for Semiconductors (ITRS): More Moore. Itrs, pp. 1–52 (2015).
4. Ionescu, A.M., Riel, H.: Tunnel field-effect transistors as energy-efficient electronic switches. Nature **479**(7373), 329–337 (2011). doi: https://doi.org/10.1038/nature10679.
5. Gopalakrishnan, K., Griffin, P.B., & Plummer, J.D.: Impact Ionization MOS (I-MOS). Part I : Dev. Circ. Simul. **52**(1), 69–76 (2005)

6. Dixit, A., Singh, S., Kondekar, P.N., Kumar, P.: Parameters optimization of Lateral impact ionization MOS (LIMOS). IEEE TechSym 2014—2014 IEEE Students' Technology Symposium, pp. 376–380 (2014). doi: https://doi.org/10.1109/TechSym.2014.6808079.
7. Lee, C.W., Afzalian, A., Akhavan, N.D., Yan, R., Ferain, I., Colinge, J.P.: Junctionless multigate field-effect transistor. Appl. Phys. Lett. **94**(5), 053511 (2009). https://doi.org/10.1063/1.3079411
8. Del Alamo, J.A.: Nanometre-scale electronics with III–V compound semiconductors. Nature **479**(7373), 317–323 (2011)
9. Del Alamo, J.A., Antoniadis, D.A., Lin, J., Lu, W., Vardi, A., Zhao, X.: Nanometer-scale III-V MOSFETs. IEEE J. Electron. Dev. Soc. **4**(5), 205–214 (2016). https://doi.org/10.1109/JEDS.2016.2571666
10. Waldron, N.: III-V Devices and technology for CMOS. In: High Mobility Materials for CMOS Applications, pp. 231–280. Woodhead Publishing (2018)
11. Dixit, A., Samajdar, D.P., Bagga, N., & Yadav, D.S.: Performance investigation of a novel GaAs1-xSbx-on-insulator (GASOI) FinFET: Role of interface trap charges and hetero dielectric. Mater. Today Commun. **26**, 101964 (2021). https://doi.org/10.1016/j.mtcomm.2020.101964
12. Dixit A., Prakash Samajdar D., Sharma D.: Performance analysis of Ga0.47In0.53Sb-FinFET and Si-FinFET for RF and low-power design applications. In: Das, N.R., Sarkar S. (eds.) Computers and Devices for Communication. CODEC 2019. Lecture Notes in Networks and Systems, vol. 147. Springer, Singapore (2021). https://doi.org/10.1007/978-981-15-8366-7_78
13. Chang, S., Zhou, H., Gong, N., Kang, D., Lim, J.: Fin-width effects on characteristics of ingass-based independent. IEEE Electron. Dev. Lett. **38**(4), 441–444 (2017)
14. Convertino, C. et al.: InGaAs FinFETs directly integrated on silicon by selective growth in oxide cavities, vol. 111 (2019). doi: https://doi.org/10.3390/ma12010087.
15. Zeng, Y., et al.: InAs FinFETs performance enhancement by superacid surface treatment. IEEE Trans. Electron. Dev. **66**(4), 1856–1861 (2019). https://doi.org/10.1109/TED.2019.2901281
16. Jang, E., Shin, S., Jeong, J.W., Kim, K.R.: Low leakage III-V/Ge CMOS FinFET design for high-performance logic applications with high-k spacer technology. J. Semicond. Technol. Sci. **18**(3), 295–300 (2018). https://doi.org/10.5573/JSTS.2018.18.3.295
17. Synopsys, T.: Sdevice Manual. Release H-2014.06, Zurich, Switzerland
18. Synopsys, T.: Sentaurus Structure Editor User Guide. Mountain View CA (2014)
19. Dixit, A., Samajdar, D.P.: Extraction of performance parameters of nanoscale SOI LDD-FinFET using a semi-analytical model of capacitance and channel potential for low-power applications. Appl. Phys. A Mater. Sci. Process. **126**(10) (2020). doi: https://doi.org/10.1007/s00339-020-03970-z.
20. Auth, C., et al.: A 22nm high performance and low-power CMOS technology featuring fully-depleted tri-gate transistors, self-aligned contacts and high density MIM capacitors. Symp. VLSI Technol. (VLSIT) **2012**, 131–132 (2012). https://doi.org/10.1109/VLSIT.2012.6242496
21. Kumar, P., Singh, S., Kondekar, P., Dixit, A.: Digital and analog performance of gate inside P-type junctionless transistor (GI-JLT). In Proceedings of International Conference on Computational Intelligence, Modelling and Simulation (2013). doi: https://doi.org/10.1109/CIMSim.2013.70.
22. Hu, H.H., Cheng, H.Y.: A hybrid wide drain Poly-Si FinTFT for RF application. IEEE Access **6**, 47268–47272 (2018). https://doi.org/10.1109/ACCESS.2018.2865881

Radiation Tolerant Memory Cell for Aerospace Applications

Pawan Kumar Sahu, Sparsh Koushik, Shashank Kumar Dubey, and Aminul Islam

1 Introduction

The dimensions and voltage level of present-day electronic circuits have reduced drastically, this decrease in operation voltage and smaller size of these devices have increased their sensitivity to radiation [1]. Such electronic systems when used in space applications require robustness against radiation. There are many effects of radiation that range from data flipping to permanent device failure. Memory circuit, particularly static random access memory, is an integral part of all space electronics equipment (i.e., Command and Data-handling Systems) that process critical data. Therefore, there is a dire need for PVT variation-aware and radiation-hardened memory circuit for aerospace applications. Radiation-induced soft errors such as single-event effect (SEE) or total ionizing dose (TID) cause a significant threat to the reliability of electronic systems. SRAM circuits are more sensitive to soft-error rate due to SEE. In modern electronic computer chips, the density of SRAM is large for high performance, resulting in an increased soft-error rate (SER) [2]. SRAM cells occupy more than half of the chip area of modern-day high-performance ICs. As CMOS technology is scaled, the spacing between the transistors is reduced. This makes multiple transistors susceptible to the charge deposited from a single particle causing single-event–multiple-node upsets [3]. This encourages the need for soft-error resilient SRAM cells that can function even under extreme radiation conditions.

SRAM is critical to the speed of any processor, as it is used as cache memory in almost all the processors in electronic equipment. The performance characteristics of SRAM cell is desired to be superior for the design of high-speed processors.

P. K. Sahu · S. Koushik · S. K. Dubey (✉) · A. Islam
Department of Electronics and Communication Engineering, Birla Institute of Technology, Mesra, Ranchi, Jharkhand, India

A. Islam
e-mail: aminulislam@bitmesra.ac.in

A. Biswas et al. (eds.), *Microelectronics, Circuits and Systems*, Lecture Notes in Electrical Engineering 755, https://doi.org/10.1007/978-981-16-1570-2_10

Radiation-hardened SRAM cells are also important for terrestrial applications. Sometimes in terrestrial applications also the packaging material of the chip may give away radiation, this radiation may cause SEU in the chip. Hence in terrestrial applications also for highly reliable circuits, we require radiation-hardened cells.

A variety of structures have been proposed that address the problem of radiation-induced soft errors. Authors in [4] have proposed two radiation-hardened-by-design (RHBD) memory cells (PS-10T and NS-10T) by using a stacked structure. However, these memory cells cannot provide fully single-event upset (SEU) protection. NS-10T cell can only recover $0 \rightarrow 1$ SEU whereas, PS-10T cell has the capability of tolerating only $1 \rightarrow 0$ SEU. In [5], the authors proposed a differential read scheme-based SRAM cell. 12T Dual Interlocked storage Cell (DICE-12T) was proposed in [6], which uses dual node feedback control to make it fully immune against SEU on a sensitive node. However, it is unable to recover single-event–multiple-node upsets (SEMNUs), which is becoming a more critical reliability issue for emerging nanometer CMOS technology. The 12-transistor radiation-hardened-by-design (RHD-12T) memory cell was proposed in [7]. Besides providing SEU immunity on any of its internal single nodes, it can also provide the SEMNUs immunity to some extent. Another RHBD-10T cell is recently proposed in [8] to provide area-efficient solution to SEU immunity. However, its radiation hardness performance in terms of SEMNUs is yet to be investigated. The previous solutions either show larger area overhead or limited SEU immunity. Therefore, they may not be suitable for aerospace applications where both area-efficient and highly reliable performances are required. In order to limit the soft error in SRAM, traditionally error-correcting codes (ECC) are used. However, due to larger area overhead and longer processing time, ECC-protected SRAM is not suitable for aerospace applications. Hence radiation hardness by design is preferred for enhanced performance.

The remaining part of the paper has been organized as follows. Section 2 explains the proposed SPS-10T memory cell, SEU recovery analysis, SEMNU recovery analysis, MEMNU recovery analysis, and the implantation. Section 3 explains the evaluation of the different parameters of the proposed cells and comparison cells. This includes access time evaluation and comparisons, cost comparison, SEU robustness verification and comparison, stability comparison (read static noise margin (RSNM), write margin (WM), and hold static noise margin (HSNM)), read power comparison, write power comparison, and hold power comparison. Section 4 contains the conclusion.

2 Proposed SPS-10T Memory Cell

2.1 Cell Structure and Behavior

The SPS-10T memory cell has four storage nodes VL, Q, QN, and VR. Q and QN are the output storage nodes, and VL and VR are the intermediate storage nodes. The

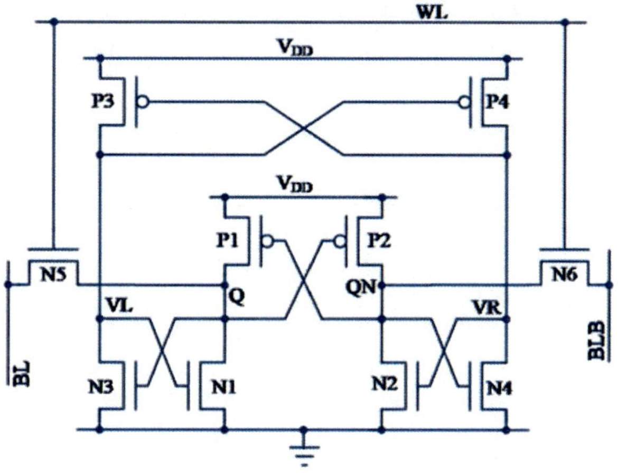

Fig. 1 Proposed SPS-10T memory cell

transistors N5 and N6 are the access transistors which are connected with the storage nodes Q and QN, also connected with the bit lines BL and BLB, respectively. The access transistor is controlled by the word line WL. If the stored node Q is at zero state for the proposed SPS-10T cell as shown in Fig. 1, then the logic state of QN, VL, and VR is 1, 1, and 0, respectively.

If the stored node QN is at zero state for the proposed SPS-10T cell, then the logic state of Q, VL, and VR nodes is at 1, 0, and 1, respectively. This is the complete scenario for the stable state of this proposed SPS-10T memory cell. The NMOS transistors N3 and N1 or N2 and N4 are cross-coupled to pass the output storage node value to the feedback path, and the PMOS transistors P3 and P4 are also cross-coupled and it is used as a positive feedback which helps to recover the storage node value at Q and QN. For the case when Q is at logic-1.

(1) In hold mode, the logic state of the word line (WL) is set to be ‘0’ and the two access transistors N5 and N6 are disabled, and transistors N3, P4, N2, and P1 are turned ON while N1, P3, N4, and P2 are turned OFF. In this mode, the storage nodes Q and QN are assumed to hold their values as 1 and 0, respectively.
(2) Prior to the read operation, the bit lines BL and BLB are pre-charged to supply voltage to reduce the wakeup time and then for the read operation, WL is set to be ‘1’ state to enable the access transistors N5 and N6. As for the ‘read 1’ operation, BL will maintain its original state without being discharged, but precharged bit line BLB will discharge through N6 and N2 to GND. While discharging N6 is in saturation and N2 is in the triode region. The voltage at BLB starts to decrease and the difference of the voltage between BL and BLB will be amplified by a differential sense amplifier (not shown) to decipher the stored value in the memory cell. To obtain one proper read operation (WN1/LN1)/(WN5/LN5) or (WN2/LN2)/(WN6/LN6) should be properly sized to get good RSNM. Usually (WN1/LN1)/(WN5/LN5) or

(WN2/LN2)/(WN6/LN6) is greater than 1 so that the voltage at QN should not exceed the value $V_{QN} \leq V_{DD} - |V_{thp}|$ and due to less resistance of N1 and N2, we are able to get good RSNM.

During the write operation at Q and QN of the SPS-10T memory cell, WL and BLB are set to '1' while BL is set to be '0'. The charged node Q finds a discharging path through N5. BL pulls down the potential at output storage node Q below the threshold voltage of N3 resulting in a switch of the transistor N3, meanwhile transistor P2 is ON and trying to pull up the node QN, at the same time N6 also tries to pull up the potential at node QN. Transistors P2 and N6 are trying to pull up the node QN and N2 is trying to pull it down but the pull strength due to P2 and N6 are in parallel is more than the pull-down strength due to N2. N2 becomes almost OFF because as VQ decreases through N5 to grounded BL and hence N3 stops conducting and VL rises putting P4 OFF and lowering VR. The lower potential of VR puts N2 OFF. The pull-up ratio (WP1/LP1)/(WN5/LN5) or (WP2/LP2)/(WN6/LN6) is less than 1 even for the same size transistors, considering the PMOS transistor has small carrier mobility than NMOS carrier mobility. For higher writeability or for a good write operation, the strength of the access transistor should be greater than that of pull-up transistor. The write operation is completed when the content is flipped. That is, after the write operation, the values of the potential of storage nodes Q and QN are 0 and 1. While the potentials of intermediate nodes VL and VR are 1 and 0, respectively. Strong 0 and 1 at Q and QN are obtained with the help of cross-coupled pull-down NMOS transistors N3 and N4 and feedback pull-up transistors P3 and P4.

2.2 *SEU Recovery Analysis*

Taking the case when the state of the sensitive nodes Q, QN, VL, and VR is 0, 1, 1, and 0, respectively:

(1) When the output storage node Q is affected by an SEU, and its potential changes from 0 to 1, transistor P2 is OFF and transistor N3 is ON momentarily, as P2 is OFF, QN is at a high impedance state and its state is kept at logic-1. Meanwhile N3 is ON and P3 is already ON making the VL node unstable for a while and its potential is decided by the strength of P3 and N3 as we kept the strength of P3 more than the strength of N3, so the value of node VL becomes 1 and it will switch ON the transistor N1, which will help to recover the value at node Q immediately.

(2) When the output storage node QN is affected by an SEU and its potential changes from 1 to 0, transistor N4 is OFF and transistor P1 is ON temporarily. As N4 is OFF, node VR is at a high impedance state and its state is kept at logic-0. Meanwhile, P1 is ON and N1 is already ON which will make node Q unstable for a while and its potential is decided by the strength of transistors P1 and N1, as we kept the strength of N1 higher than that of P1. So the logic is decided by the pull-down transistor N1 and the value at node Q is 0. It will

switch ON the transistor P2 again and the node voltage of QN is recovered back to logic-1.

2.3 MEMNU Recovery Analysis

Multiple events $0 \rightarrow 1$ and $1 \rightarrow 0$ can occur at multiple nodes. Nodes Q and QN are capable of handling both 0 to 1 and 1 to 0 multiple events simultaneously. First, we take the case, when the state of nodes Q, QN, VL, and VR is 0, 1, 1, and 0, respectively.

When the nodes Q and QN change from $0 \rightarrow 1$ and $1 \rightarrow 0$, respectively, transistor P2 is OFF and transistor N3 is ON momentarily, transistor N4 is OFF and transistor P1 is ON. As transistor N4 is OFF, VR is at a high impedance state and it is kept at logic-0. Meanwhile, N3 is ON and P3 is already ON making the VL node unstable and its potential is decided by the strength of P3 and N3, as we kept the strength of P3 more than the strength of N3. So the value at node VL becomes 1 and it will again switch ON the transistor N1 and try to pull down the potential at Q to zero but P1 is also ON due to upset at QN from $1 \rightarrow 0$, so the potential at node Q is now decided by the strength of P1 and N1 and as we kept the strength of N1 more than the strength of P1, node value of Q becomes 0 and it will switch ON the PMOS transistor P2 which will help to recover the value at node QN. In this way with the help of feedback transmitters P3 and P4, the node voltages of Q and QN are recovered very fast. Thus, the proposed SPS-10T is capable enough to handle simultaneous upset at both the output-sensitive nodes Q and QN.

2.4 Implementation

The proposed SPS-10T memory cell operated at nominal voltage 1.2 V. This cell is implemented in 22-nm CMOS Technology. The layout of the proposed cell SPS-10T is shown in Fig. 2. The cell ratio $(WN_1/LN_1)/(WN_5/LN_5)$ or $(WN_2/LN_2)/(WN_6/LN_6)$ is set to be 2.4 for better read stability. The pull-up ratio $(WP_1/LP_1)/(WN_5/LN_5)$ or $(WP_2/LP_2)/(WN_6/LN_6)$ is set to be 1.2. The second pull-up ratio $(WP_3/LP_3)/(WN_3/LN_3)$ or $(WP_4/LP_4)/(WN_4/LN_4)$ is set to be 4.5 for providing better feedback and easy recovery of the storage nodes Q and QN. All the simulations are performed using SPICE. Double exponential current is also modeled in the SPICE which mimics the same upsets as occurred in the SPACE.

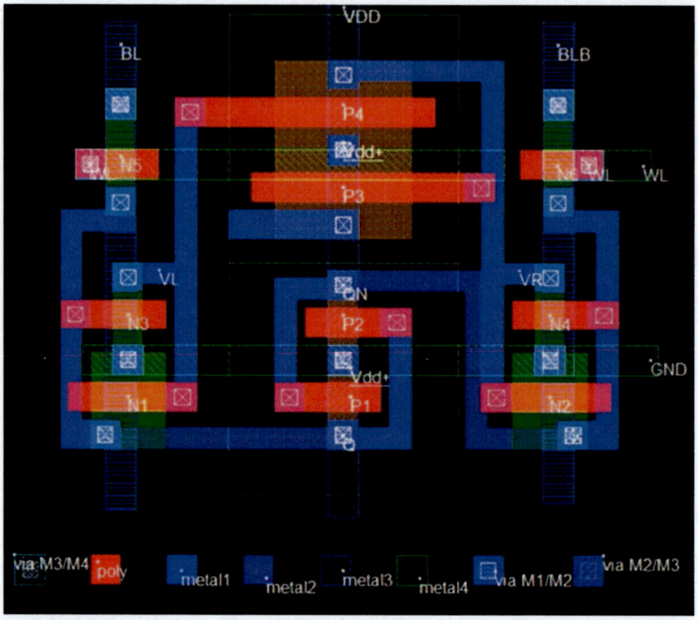

Fig. 2 Layout of the proposed SPS-10T memory cell

3 Simulation Results and Discussion

3.1 Access Time Comparison

The read access time (T_{RA}) is measured as the time taken for the voltage of BLB line to fall by 50 mV in the read condition, and this time measurement starts from the time when WL starts to be asserted. The simulation setup for the read condition is as follows: the BL and BLB line are pre-charged to 1, the storage nodes Q and QN are pre-charged to 1 and 0, respectively, and the WL is swept from 0 to 1. This value of 50-mV is chosen because this is the typical value required by the sense amplifier to conclusively detect a level. The read access time depends on the access transistors. More the current through access transistors, the lesser the access time. This current is dependent on the sizing of the access transistor and the pull-down transistor. Bit-line loading also affects the read access time.

Write access time is measured as the time taken for the voltage of QN node of the given SRAM to reach 90% of the V_{DD} in the write condition, and this time measurement starts from the time when WL starts to be asserted. The simulation condition for write condition is as follows: the BL is connected to GND and BLB line is connected to V_{DD}, the storage nodes Q and QN are pre-charged to 1 and 0, respectively, and the WL is swept from 0 to 1.

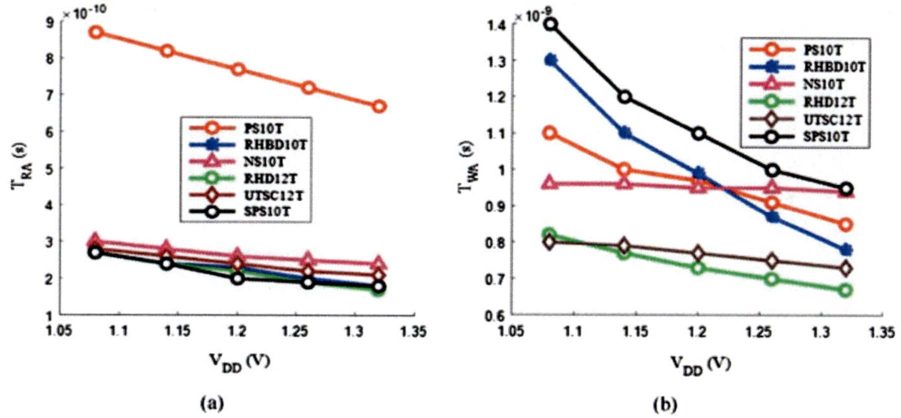

Fig. 3 Plot of **a** T_{RA} and **b** T_{WA} of proposed and comparison cells with V_{DD}

The T_{RA} and T_{WA} for the proposed structure were calculated at the supply voltages of 1.08, 1.12, 1.2, 1.26, and 1.32 V. The supply voltage of design is 1.2 V. The other values of supply voltage used are to account for supply voltage variation by 5 and 10%. Similarly, T_{RA} and T_{WA} have been calculated for all the comparison structures namely 6T, PS-10T, RHBD-10T, NS-10T, RHD-12T, and UTSC-12T at these five voltage values. The values of T_{RA} and T_{WA} of all these structures at the different supply voltages can be seen in Fig. 3a and b, respectively.

The T_{RA} as shown in Fig. 3a increases for all the structures as the supply voltage decreases. This can be easily explained as lower supply voltage means lower current and hence larger time to discharge the bit-line capacitance. The T_{RA} of the proposed cell is found to be 200 ps at the proposed supply voltage of 1.2 V. The T_{RA} for the comparison cells are as follows: 200 ps for 6T, 770 ps for PS-10T, 230 ps for RHBD-10T, 260 ps for NS-10T, 220 ps for RHD-12T, and 240 ps for UTSC-12T. It can be seen that the T_{RA} of the proposed cell was found to be the shortest among the structures that support single-event upset resistance. It can be noticed that the proposed SPS-10T has comparable T_{RA} with RHBD-10T, NS-10T, RHD-12T, and UTSC-12T.

The T_{WA} as shown in Fig. 3b increases for all structures as the supply voltage decreases. The reason is the same as that for T_{RA}. The T_{WA} of the proposed cell is found to be 1100 ps at the supply voltage of 1.2 V. The point to note is that the T_{WA} is higher for the proposed cell as compared to the existing cells. This is due to a longer feedback path. A higher T_{WA} also signifies the most robustness against data flipping due to radiation. This can be considered as a tradeoff between the radiation tolerating capacity and T_{WA}.

3.2 SEU Robustness Comparison

SEU robustness was verified by using a double exponential current source. A double exponential current source was injected at the sensitive nodes, and this was done to simulate the radiation strike or single-event upset at these nodes. These sensitive nodes are characterized as the nodes which are reverse biased [1].

To check the relative stability of the different structures to a single event, we checked the maximum current the circuit can tolerate before the stored data changes. The maximum value of the peak current for which the proposed SPS-10T operated correctly is 0.956 mA. This maximum current tolerance for the different comparison cells is as follows: 72.8 μA for PS-10T, 110 μA for RHBD-10T, 82.45 μA for NS-10T, 449 μA for RHD-12T, and 71.1 μA for UTSC-12T.

3.3 Stability Comparison

The RSNM, WM, and HSNM for all the cells at the five voltages have been shown in Fig. 4a, b, and Fig. 5, respectively. The figure also contains the variation of the RSNM, WM, and HSNM with voltage. It can be observed that the RSNM increases with a decrease in voltage. This can be attributed to the low current, since lower current makes the data flipping more difficult. The RSNM and HSNM were calculated by using the butterfly plot of the SRAM cell [9].

The RSNM for the proposed SPS-10T was found to be 99 mV at the supply voltage of 1.2 V. The RSNM of the comparison cells as obtained at 1.2 V supply voltage is as follows: 40 mV for 6T, 80 mV for PS-10T, 98 mV for RHBD-10T, 75 mV for NS-10T, 100 mV for RHD-12T, and 98 mV for UTSC-12T. In comparison to other cells, the read static noise margin of the proposed cell was found to be the maximum

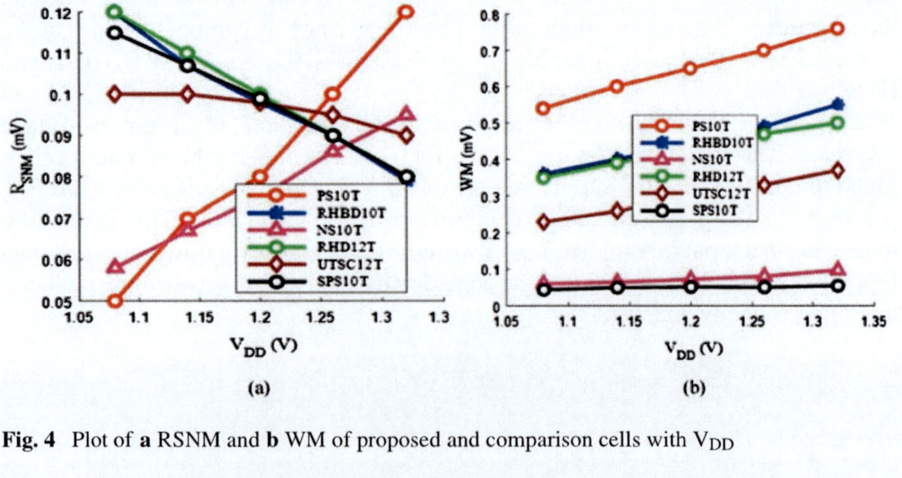

Fig. 4 Plot of **a** RSNM and **b** WM of proposed and comparison cells with V_{DD}

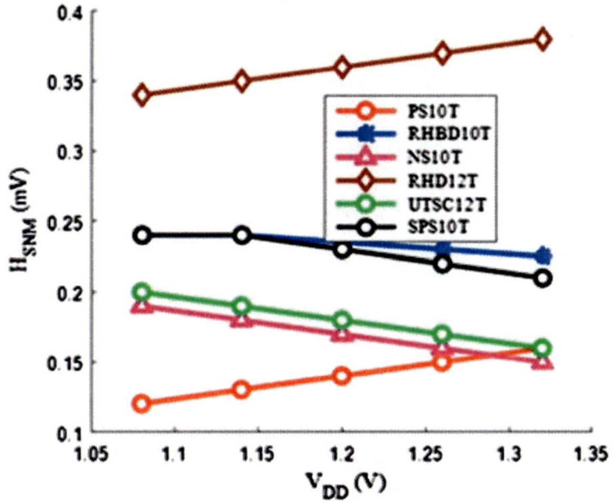

Fig. 5 Plot of HSNM of proposed and comparison cells with V_{DD}

for all of the comparison cells except RHD-12T, and the RSNM of RHD-12T is high by a very small margin of 1 mV. Hence it can be concluded that the RSNM of the proposed SPS-10T SRAM cell is superior or comparable to all the comparison cells.

The HSNM for the proposed cell SPS-10T was found to be 230 mV. The HSNM of the comparison cells as obtained at 1.2 V supply voltage is as following: 230 mV for 6T, 140 mV for PS-10T, 235 mV for RHBD-10T, 170 mV for NS-10T, 360 mV for RHD-12T, and 180 mV for UTSC-12T. As it can be observed, the HSNM for the proposed cell is not the best among the comparison cells but it is comparable to the values of the best performing circuit.

The write ability is a measure of the ability of the SRAM to undergo a write operation, i.e., data being flipped in the stored nodes. Although traditionally write static noise margin (WSNM) was used for this purpose, however, recent studies have shown that WM provides a more reliable measure of an SRAM cell's write ability [10]. By definition, WM is measured as the voltage difference between V_{DD} and WL when the storage nodes 'Q' and 'QN' flip [10].

Figure 4b shows the WM of the comparison cell and the proposed cell at the different supply voltages. The WM of the proposed cell was found to be lower than the comparison cell. This is attributed to the same logic as that for the proposed cell getting a poor T_{WA}. The stronger the feedback and the stronger the radiation tolerance, the difficult it is to write in that cell.

4 Conclusion

This paper presents a highly radiation-tolerant 10-transistor SRAM cell. The radiation tolerating capacity of the proposed cell is very high as compared to any other existing cell. The minimum charge required to change the data of the proposed cell is significantly higher than any other previously proposed cells. The area of the proposed cell is similar to most existing cells. The power requirement of the cell is also reasonably low and comparable to other cells. The read static noise margin of the transistor is superior to almost all preexisting structures. The hold static noise margin is also similar to comparison cells. Although the write ability is not good but this is a tradeoff between higher radiation tolerance and write ability. Hence we can conclude that the proposed SPS-10T cell is suitable for a very high radiation environment where there is a possibility of tolerance in write access time.

References

1. Baumann, R.C.: Radiation-induced soft errors in advanced semiconductor technologies. IEEE Trans. Device Mater. Reliab. **5**(3), 305–316 (2005)
2. Seo, D., Han, J.W., Kim, J., Chang, I.J.: A 28mn FD-SOI 4KB radiation-hardened 12T SRAM macro with 0.6~1V wide dynamic voltage scaling for space applications. In: 2018 IEEE Asian Solid-State Circuits Conference (A-SSCC), pp. 133–134. Tainan (2018)
3. Peng, C., et al.: Radiation-hardened 14T SRAM bitcell with speed and power optimized for space application. IEEE Trans. Very Large Scale Integr. (VLSI) Syst. **27**(2), 407–415 (2019)
4. Lin, D., et al.: A novel SEU tolerant memory cell for space applications. IEICE Electron. Express **15**(17), 1–8 (2018)
5. Jahinuzzaman, S.M., Rennie, D.J., Sachdev, M.: A soft error tolerant 10T SRAM bit-cell with differential read capability. IEEE Trans. Nucl. Sci. **56**(6), 3768–3773 (2009)
6. Calin, T., Nicolaidis, M., Velazco, R.: Upset hardened memory design for submicron CMOS technology. IEEE Trans. Nucl. Sci. **43**(6), 2874–2878 (1996)
7. Qi, C., Xiao, L., Wang, T., Li, J., Li, L.: A highly reliable memory cell design combined with layout-level approach to tolerant single-event upsets. IEEE Trans. Device Mater. Reliab. **16**(3), 388–395 (2016)
8. Guo, J., et al.: Design of area-efficient and highly reliable RHBD 10T memory cell for aerospace applications. IEEE Trans. Very Large Scale Integr. (VLSI) Syst. **26**(5), 991–994 (2018)
9. Islam, A., Hasan, M.: A technique to mitigate impact of process, voltage and temperature variations on design metrics of sram cell. Microelectron. Reliab. **52**(2), 405–411 (2012)
10. Ahmad, S., Gupta, M. K., Alam, N., Hasan, M.: Single-ended Schmitt-trigger-based robust low-power SRAM cell. IEEE Trans. Very Large Scale Integr. (VLSI) Syst. **24**(8), 2634–2642 (2016).

Memory Elements and Circuits

A Highly Reliable and Radiation-Hardened Majority PFET-Based 10T SRAM Cell

Aakansha, G. S. Namith, A. Dinesh, A. Sai Ram, Shashank Kumar Dubey, and Aminul Islam

1 Introduction

The aerospace market in India is estimated to reach around $70 billion by 2030. One of the major components used in it is memory. SRAM cell is used for cache memory, which occupies 90% of the chip area. The smallest transistors that are sensitive to process variations are required. The PVT (Process, Voltage, Temperature) variations affect the device threshold voltage (V_t), which in turn modifies the drain to source current (I_{DS}) [1]. The cells in the SRAM during the standby mode are inactive and hold power is consumed for the retention of data due to various leakage components. With the minimization of the technology, the memory cells of SRAM (Static Random Access Memory) have lower supply voltage and smaller node capacitance, which makes these memory cells very highly susceptible to different kinds of radiation particles, which are usually alpha particles, heavy ion and cosmic rays. Hence, these nanoscale integrated circuits become largely susceptible to such particle-induced single event transients (SETs) mainly because of the less signal charge and reduced noise margin. Primarily, SETs are caused by alpha particles and cosmic neutrons. Packaging materials and intergalactic rays are responsible for the origin of these alpha particles and cosmic neutrons respectively. Extra charge in these particles is generated through direct or indirect ionization in silicon. The generated extra charge gets collected by sensitive nodes, which is responsible for the creation of voltage transients at these sensitive nodes [2–4].

A latch consists of a cross-coupled inverter pair. The larger values of the amplitude and duration of the transient alter the value that is stored in the latch, which causes a single event upset (SEU). Since the damage done to the device is not permanent;

Aakansha · G. S. Namith · A. Dinesh · A. S. Ram · S. K. Dubey · A. Islam (✉)
Department of Electronics and Communication Engineering, Birla Institute of Technology, Mesra, Ranchi, Jharkhand, India
e-mail: aminulislam@bitmesra.ac.in

A. Biswas et al. (eds.), *Microelectronics, Circuits and Systems*, Lecture Notes in Electrical Engineering 755, https://doi.org/10.1007/978-981-16-1570-2_11

SEU is also usually referred to as a 'soft error'. However, system malfunctions can be caused by these soft errors, which can disturb the state of SRAM [5].

Three types of failures are common in SRAM cell [6]—hard failures, soft failures and parametric failures. (1) Open or short conditions cause hard failures. (2) Soft failures—Three major sources are the primary reasons for soft failures. The first major source for the soft failures is alpha particles, which get released from radioactive impurities like packaging materials [7]. The second source is the high-energy neutrons, which originate from the terrestrial cosmic radiations and due to the interaction with cosmic ray thermal neutron [5]. Parametric failure such as the variation in the design parameters is the third and one of the major sources for these failures.

In this paper, we proposed a 10T radiation-hardened SRAM cell with PMOS access transistors, which is radiation-hardened when compared with Quatro 10T SRAM cell. The organization of the paper is made in such a way that Sect. 2 presents the prior work of Quatro-10T cell, Sect. 3 describes the proposed 10T SRAM cell with PMOS access transistors, Sect. 4 analyzes the operations of the proposed 10T cell such as Read Access Time, Hold Power, Read Static Noise Margin, Write Static Margin and Soft Error Robustness of the proposed 10T cell while Sect. 5 draws the conclusion.

2 Prior Work

Figure 1 shows the basic Quatro-10T SRAM cell in which four PMOS and four NMOS transistors named as MP1, MP2, MP3, MP4 and MN1, MN2, MN3, MN4, respectively. Bit lines BL and BLB are connected with the access transistors MN5 and MN6 to the storage nodes A and B [8]. If the bit that is stored at node A is assumed to be at logic '0', then the logic values at nodes B, C and D are '1', '1' and '0' respectively. Access transistors MN5 and MN6 are cut off to show how the storage nodes maintain the same logic values when the circuit is in HOLD mode. Storage node A is having a logic value of '0', because the path of charging to VDD by the MP1 is cut off and the path to discharge to ground by the MN1 is available. This implies that PMOS transistor MP1 is OFF and NMOS transistor MN1 is ON. Gate of MN1 is connected to node B. Since the logic value at node B is maintained at '1', it switches MN1 ON, which allows node A to have the path to discharge to logic '0'. Since node B is at logic '1', MN4 is ON pulling node D down to logic '0', which turns MP2 ON, raising node C to logic '1'. Gate of MP1 is connected to node C. Hence, the logic value at node C is maintained at '1' to cutoff MP1. As the storage node C is at logic '1', and then the PMOS transistor MP4 will be OFF as the gate is connected to the storage node C. This removes the path of charging to V_{DD} for node D. MN4 is ON, as the gate is connected to storage node B, which is having a logic value of '1'. Hence the node D will discharge to '0' since it has a path to discharge to the ground.

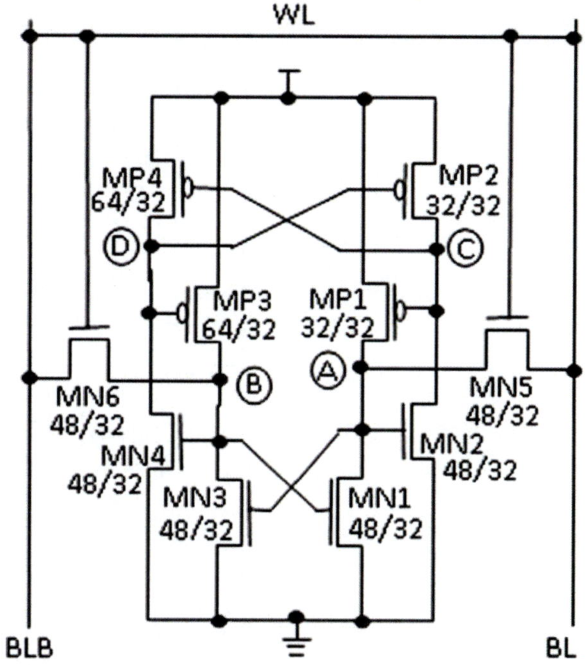

Fig. 1 Schematic of Quatro-10T SRAM cell

3 Proposed 10T SRAM Cell

This paper proposes a radiation-hardened 10T SRAM cell. As shown in Fig. 2, it includes four PMOS and four NMOS transistors named as MP1, MP2, MP3, MP4 and MN1, MN2, MN3, MN4, respectively. The two access transistors MN5 and MN6 are connected with bit lines BL and BLB to the storage nodes A and B. If the stored bit at node A is at logic '0', then the logic values at nodes B, C and D are '1', '1' and '0', respectively. The restoring of logic values at nodes happens in a similar way to the Quatro cell. Higher degree of radiation tolerance is the advantage that is obtained by using PMOS transistors as major devices. Radiation bombardment causes to increase the leakage currents in the NMOS transistors, whereas PMOS device leakage currents are not affected. The data that are stored in the cell's storage node may get corrupted due to the excessive leakage in access transistors. Hence this advantage of PMOS access transistors over NMOS access transistors will help in improving the failure probability of the SRAM cell during hold state. Smaller magnitudes of gate leakage in PMOS devices over NMOS devices also add as a beneficial factor to the cell [6].

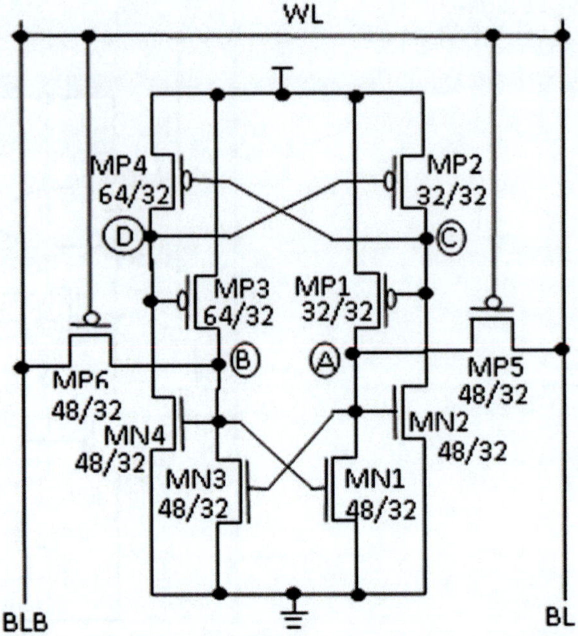

Fig. 2 Schematic of the proposed 10T SRAM cell with PMOS access transistors

4 Operations of the Proposed 10T SRAM Cell

4.1 Read Access Time (T_{RA})

The read access time (T_{RA}) is observed starting at the instant when word line (WL) gets activated to the instant when bit line or bit line bar gets discharged by a value of 50 mV from its initial high level. 50 mV value is sufficient to be detected by sense amplifier [9].

In Quatro 10T cell, the access transistors are NMOS whereas, in the proposed 10T cell, the access transistors are PMOS. The majority of charge carriers in NMOS are electrons whereas the majority of charge carriers in PMOS are holes. As we know that electrons are more mobile as compared to holes, it takes less read time. This explains why the T_{RA} of the proposed cell is marginally longer compared to that of Quatro 10T cell as shown in Fig. 3a. The Monte Carlo analysis is done with 5000 sample size and the distribution plot is shown in Fig. 3b for both Quatro and proposed cells. We observed a greater mean in Quatro as compared to the proposed cell. The spread of the Gaussian-like curve is better for the proposed cell.

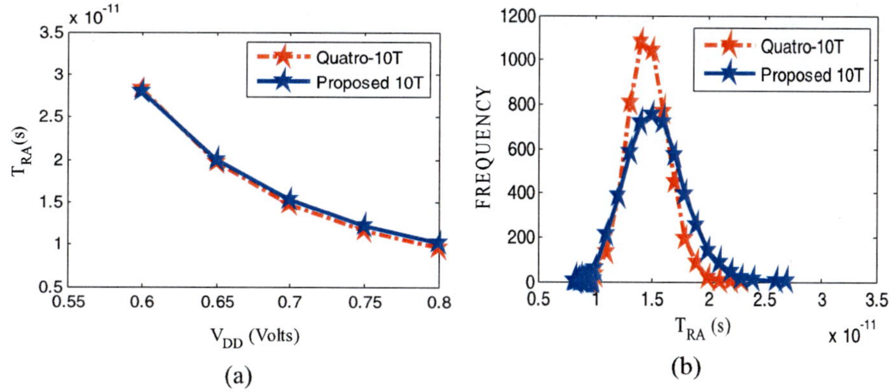

Fig. 3 **a** T_{RA} of Quatro 10T cell and proposed 10T cell at various V_{DD}. **b** Comparison of T_{RA} distribution plots for Quatro-10T and proposed 10T SRAM cells

4.2 Hold Operation

Hold operation is important for data retention in case of high leakage current. For the purpose of long data retention during hold mode, disabling of word line (WL) is done and the BL and BLB lines are made precharged and tight connection of the partial cross-coupled inverters is established [10]. We measured the hold power using transient analysis by taking the readings of $I_{avg.}$ and we calculated the hold power by multiplying the value of $I_{avg.}$ with the supply voltage. In the case of the proposed cell, WL is made high to cut off the access transistors for hold operation and the rest calculation remains the same. In the case of hold operation, the access transistors are cut off so there is no change in the operation for the Quatro and proposed cells. Hence, there is no change in the value of hold power. Figure 4 shows that as we keep on increasing the values of V_{DD}, the value of hold power also increases gradually.

Fig. 4 Hold power of Quatro 10T cell and proposed 10T cell at various V_{DD}

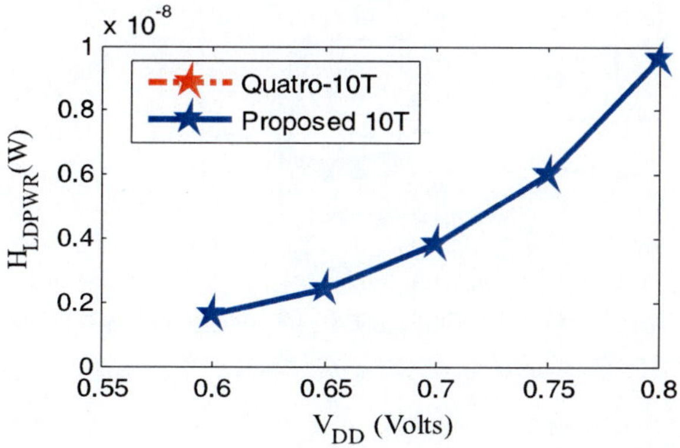

4.3 Read Static Noise Margin (RSNM) and Write Static Noise Margin (WSNM)

The minimum DC noise voltage which is needed to flip the cell is known as Static Noise Margin (SNM) [11]. It is one of the rapidly used design parameters to measure the stability of the cell. We can observe from Fig. 5a that the butterfly curve for both Quatro 10T and the proposed 10T is obtained in a similar manner. This is because, for a stable read condition, the DC noises are affecting the transistors, which have the capacity to drive the storage nodes but not the access transistors. So, there will be no change in the value of RSNM for both Quatro and proposed 10T cells. Therefore, both the butterfly curves overlap each other as shown in Fig. 5a. The square is placed in the smaller lobe and the RSNM value is obtained to be 222 mV.

The capability of the cell to pull down the node which has been storing high value "1" to a voltage that is below the value of V_M which is the switching threshold of the other inverter which has been storing a low value of "0" contributing to the flipping of the cell is called the write static noise margin (WSNM) [12, 13]. The VTC curves are obtained as shown in Fig. 5b. We observe that the curves are overlapping for both Quatro and proposed 10T cells, i.e., they are having the same values of WSNM. This is because the access transistors are unaffected so the value of WSNM for Quatro and proposed 10T cells is the same. The value of WSNM is measured by inscribing the smallest square inside lower half of the curve and the respective smallest square's side length provides us with the value of WSNM [13], which is equal to 175 mV for both Quarto and proposed 10T cells.

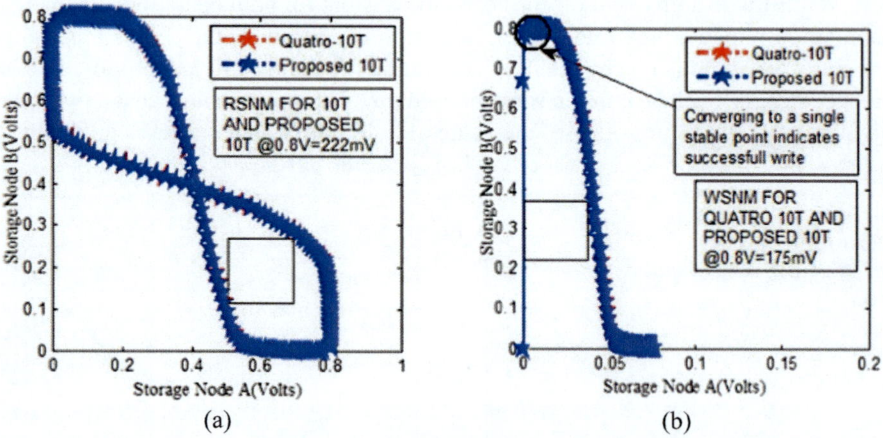

Fig. 5 **a** RSNM butterfly curves for Quatro 10T and proposed cell. **b** WSNM of Quatro and proposed 10T cell

4.4 Soft Error Robustness

For the Quatro and proposed 10T cell, the soft error robustness is verified by injecting an exponential current at each of the nodes A, B, C and D to mimic or recreate a particle-induced SET [14–18]. The Quatro 10T and the proposed 10T are capable of recovering from 1 → 0 as shown by the results in Figs. 6a and 7a, respectively, for nodes A and B and Figs. 6b and 7b for nodes C and D. Figure 6a illustrates that the capability of recovering from 1 → 0 SET, which is introduced at node A.

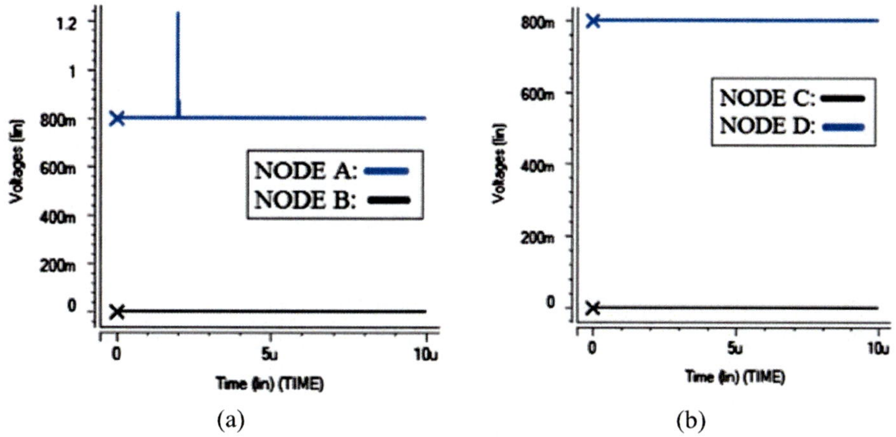

Fig. 6 Recovery of Quatro-10T cell after injecting an exponential current pulse (pulse width of 50 ns and peak voltage selected based on current margin) a 1 to 0 at nodes: **a** A and B (for all values of current), **b** C and D (for all values of current)

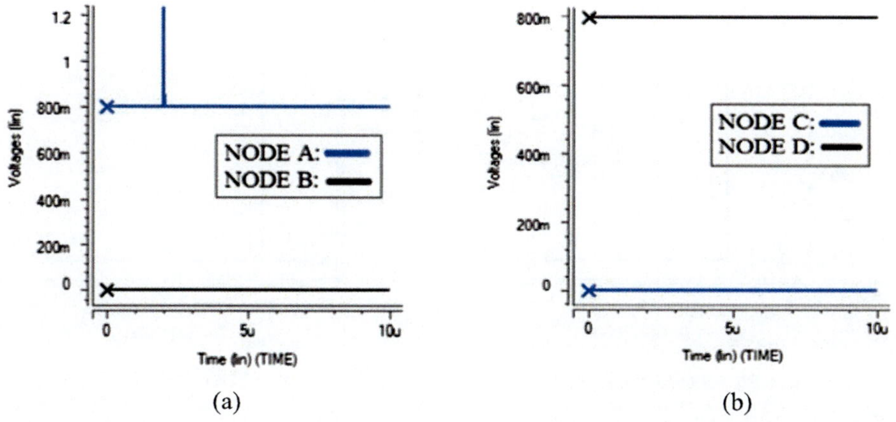

Fig. 7 Recovery of the proposed-10T cell after injecting an exponential current pulse (pulse width of 50 ns and peak voltage selected based on current margin) a 1 to 0 at nodes: **a** A and B (for all values of current), **b** C and D (for all values of current)

As, we can see from Fig. 6a, node A after getting affected by the exponential current pulse is able to recover to 1. The sudden rise from 1 is due to the exponential pulse, which we introduced at node A. The other nodes also seem unaffected from Fig. 6b. The proposed 10T also shows a similar behavior such as the Quatro 10T cell, which can successfully recover if 1 → 0 SET is introduced at node A. Both figures Fig. 7a, b seem similar due to the similar exponential pulse introduced at both places and the same supply voltage is taken in both cases, but the nodes at which recovery occurred is different. Quatro and proposed 10T cell has the capability to recover from 1 → 0 SET at node A whereas a sufficiently large or strong 0 → 1 SET at node A can flip the cell. When a large 0 → 1 SET is occurring at node A $V_A = 0$, then N2 has the capability to turn the transistors P1 and P4 ON by reducing the value of V_C whereas N3 has the capability to turn OFF N4 by lowering the value of V_B, in this way the cell is flipped. But the proposed 10T is made more robust against the 0 → 1 SET compared to that of Quatro 10T by replacing the NMOS access transistors with the PMOS access transistors in the proposed cell. The cell becomes more radiation hardened by using PMOS access transistors because the leakage currents in the PMOS access transistors do not get affected by the radiation bombardment whereas the leakage currents in NMOS access transistors increase rapidly, which corrupts the stored data.

The current in the exponential current spike reaches 230 µA, we can see that the Quatro 10T cell is flipped as shown in Fig. 8a for nodes A and B and Fig. 8b for nodes C and D. From the above Fig. 8a, b, we can see that the cell got flipped when 0 → 1 exponential current spike reached the value of 230 µA.

From Fig. 9a, b, we can see that the cell gets flipped when the value of current of the exponential current spike reaches 233 µA. Here, we can observe that we have an increase in the value of current margin by 3 µA in the case of the proposed cell

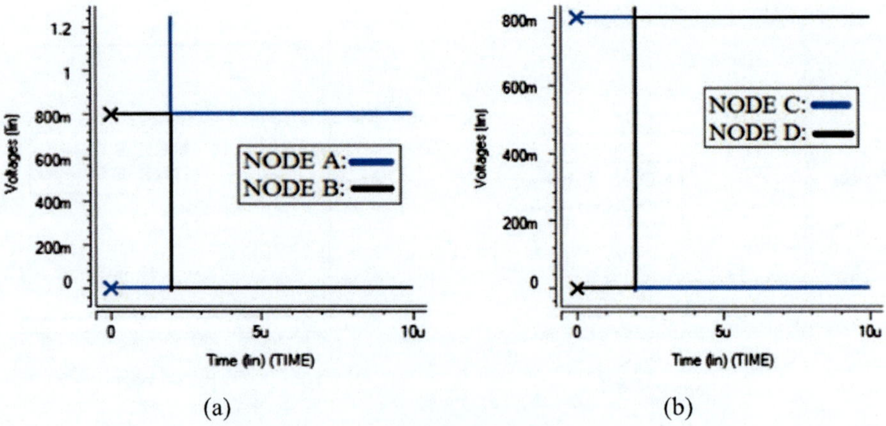

Fig. 8 Non-recovery of the Quatro-10T cell after injecting an exponential current pulse (pulse width of 50 ns and peak voltage of 230 µA) 0 to 1 at nodes: **a** A and B (for all values of current), **b** C and D (for all values of current)

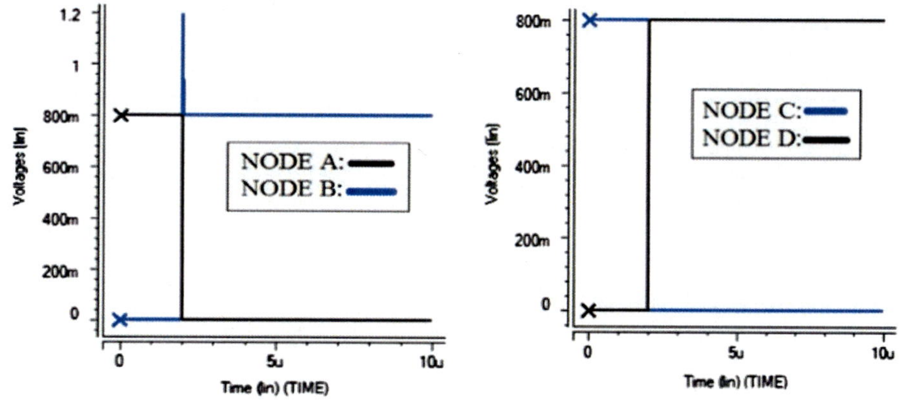

(a) (b)

Fig. 9 Non-recovery of the proposed-10T cell after injecting an exponential current (pulse width of 50 ns and peak voltage of 233 μA) 0 to 1 at nodes: **a** A and B (for all values of current), **b** C and D (for all values of current)

compared to that of Quatro cell. Therefore, the proposed 10T cell can withstand 3 μA stronger 0 → 1 SET compared to that of Quatro cell. This makes the proposed cell more robust for the soft error than the Quatro cell. The reasons for this robustness are mentioned in the above discussion.

5 Conclusion

A 10-T SRAM cell has been presented that uses PMOS access transistors, which is more soft error robust, compared to that of Quatro-10T cell at 22-nm technology. All the readings are taken with 5000 sample size while performing Monte Carlo simulation. The increment of 3 μA in current margin implies that the proposed cell withstands flipping to current spikes, which are stronger, which makes it reliable compared to Quatro-10T cell. The unaffected PMOS transistors leakage currents after radiation bombardment and hence usage of PMOS access transistors in proposed 10-T SRAM cell makes it more radiation-hardened than Quatro-10T cell. The increment in the current margin suggests the better application of the proposed cell in the radiation environment or in the aerospace industry. Therefore, the proposed cell remains as an attractive choice for aerospace applications.

References

1. Islam, A., Hasan, Md.: Variability analysis of 6T and 7T SRAM cell in sub-45nm technology **12**, 13–30 (2011)
2. Dhillon, Y.S., Diril, A.U., Chatterjee, A., Singh, A.D.: Analysis and optimization of nanometer CMOS circuits for soft-error tolerance. IEEE Trans. Very Large Scale Integr. (VLSI) Syst. **14**(5), 514–524 (2006)
3. Baumann, R.: Soft errors in advanced computer systems. IEEE Des. Test Comput. **22**(3), 258–266 (2005)
4. Dodd, P.E., Massengill, L.W.: Basic mechanisms and modeling of single-event upset in digital microelectronics. IEEE Trans. Nucl. Sci. **50**(3), 583–602 (2003)
5. Krueger, D., Francom, E., Langsdorf, J.: Circuit design for voltage scaling and SER immunity on a quad-core itanium® processor. In: IEEE International Solid-State Circuits Conference—Digest of Technical Papers, pp. 94–95. San Francisco, CA (2008)
6. Zhang, W., Li, L., Hu, J.: Design techniques of p-type CMOS circuits for gate leakage reduction in deep submicron ICs. In: 52nd Midwest Symposium on Circuits and Systems—MWSCAS 2009, pp. 551–554. Cancun, Mexico (2009)
7. Baumann, R.C.: Radiation-induced soft errors in advanced semiconductor technologies. IEEE Trans. Device Mater. Reliab. **5**(3), 305–316 (2005)
8. Jahinuzzaman, S.M., Rennie, D.J., Sachdev, M.: A soft error tolerant 10T SRAM bit-cell with differential read capability. IEEE Trans. Nucl. Sci. **56**(6), 3768–3773 (2009)
9. Islam, A., Hasan, M.: Leakage characterization of 10T SRAM cell. IEEE Trans. Electron Devices **59**(3), 631–638 (2012)
10. Ahmad, S., Alam, N., Hasan, M.: Pseudo differential multi-cell upset immune robust SRAM cell for ultra-low power applications. AEU—Int. J. Electron. Commun. **83**, 366–375 (2008)
11. Jung, I., Kim, Y., Lombardi, F.: A novel sort error hardened 10T SRAM cells for low voltage operation. In: IEEE 55th International Midwest Symposium on Circuits and Systems (MWSCAS), pp. 714–717. Boise, ID, USA (2012)
12. Islam, A., Hasan, M.: Variability aware low leakage reliable SRAM cell design technique. Microelectron. Reliab. **52**(6), 1247–1252 (2012)
13. Pal, S., Islam, A.: 9-T SRAM cell for reliable ultralow-power applications and solving multibit soft-error issue. IEEE Trans. Device Mater. Reliab. **16**(2), 172–182 (2016)
14. Qi, C., Xiao, L., Wang, T., Li, J., Li, L.: A highly reliable memory cell design combined with layout-level approach to tolerant single-event upsets. IEEE Trans. Device Mater. Reliab. **16**(3), 388–395 (2016)
15. Jiang, J., Xu, Y., Zhu, W., Xiao, J., Zou, S.: Quadruple cross-coupled latch-based 10T and 12T SRAM bit-cell designs for highly reliable terrestrial applications. IEEE Trans. Circuits Syst. I Regul. Pap. **66**(3), 967–977 (2019)
16. Guo, J., et al.: Design of area-efficient and highly reliable RHBD 10T memory cell for aerospace applications. IEEE Trans. Very Large Scale Integr. (VLSI) Syst. **26**(5), 991–994 (2018)
17. Lin, D., et al.: A novel SEU tolerant memory cell for space applications. IEICE Electron. Express **15**(17), 20180656 (2018)
18. Peng, C., et al.: Radiation-hardened 14T SRAM bitcell with speed and power optimized for space application. IEEE Trans. Very Large Scale Integr. (VLSI) Syst. **27**(2), 407–415 (2019)

Design and Analysis of Ultra-Low Power Memory Architecture with MTCMOS Asymmetrical Ground-Gated 7T SRAM Cell

Vaibhav Neema, Priyanka Parihar, and Santosh Kumar Vishvakarma

1 Introduction

Static Random Access Memory (SRAM) is popularly used as the fastest memory in system design. The memory architecture of SRAM consists of many cell blocks, which are arranged in an array. In digital system design, memory arrays are an essential and basic building block. It stores an important and large amount of data in it. The memory array consists of wordlines, bitlines, SRAM cells, sense amplifiers and decoders. A wordline selects only a single row in the memory array to be read or written and only one wordline is HIGH at any given time according to the provided address by the address decoder. The main challenge for new SRAM cell designer to focus on increase the bit count (increase in memory capacity) and maintain low power dissipation and high stability. The static noise margin (SNM) is considered as an important performance parameter of SRAM circuits as it defines the reliability of the memory [1]. In this paper, MTCMOS technology [2] is proposed to design SRAM cell for minimizing leakage current [3], power dissipation and for enhancing write, read and hold SNM [4, 5]. In MTCMOS general architecture a SLEEP transistor is connect in between cell and ground node, during the non-operational condition the SLEEP transistor is put in off state which disconnect the cell with ground node that minimize the unnecessary flow of current during non-operational mode. Here in the

V. Neema (✉)
Department of Electronics and Telecommunication, Institute of Engineering and Technology-Devi, Ahilya University, Indore 452017, India
e-mail: vneema@ietdavv.edu.in

P. Parihar
Department of Electronics and Instrumentation, Institute of Engineering and Technology- Devi, Ahilya University, Indore 452017, India

S. K. Vishvakarma
Discipline of Electrical Engineering, Indian Institute of Technology, Indore 453552, MP, India
e-mail: skvishvakarma@iiti.ac.in

A. Biswas et al. (eds.), *Microelectronics, Circuits and Systems*, Lecture Notes in Electrical Engineering 755, https://doi.org/10.1007/978-981-16-1570-2_12

proposed work, a high threshold sleep transistor is used in the asymmetrical SRAM cell, which is turned OFF in non-operational mode and helps in reducing the leakage current.

In SRAM, the 6T SRAM cell is widely used in industrial applications because of better electrical characteristics compared to the other SRAM cell architectures and well-defined fabrication steps. At nanotechnology era, the dominating process variations and voltage scaling make difficult to achieve high write and read margins in standard 6T SRAM cell. In conventional 6T SRAM cell [6], as the data read operation is directly performed through the bitlines, the data read stability gets reduced. To achieve the higher stability with less leakage current, power dissipation asymmetrical ground-gated 7T SRAM [7] cell is proposed in this paper.

As to improve the read and write stability in the asymmetrical SRAM cells, separate read and write bitlines are used to read and write the data from/into the memory respectively. The separate read bitline prevents the data flip during the read operation and stored data remain in the cell.

The paper has been organized in the following pattern. Simulation parameters are discussed in Sect. 2. Characterization of asymmetrical ground-gated 7T SRAM cell is discussed in Sect. 3. Characterization of memory array and architecture is discussed in Sect. 4. The conclusion of the paper is discussed in Sect. 5.

2 Simulation Parameters

For performance evaluation and compression of four important memory cell metrics, parameters are considered in this work.

The considered major metric parameters are

i. Stability.
ii. Memory cell delay
iii. Leakage current
iv. Static power dissipation.

2.1 Data Stability

The Static Noise Margin (SNM) [5, 8] is the most important metric parameter to define the stability of the memory cell during the write, read and hold operations in memory. The higher value of SNM define that the memory is more immune to noise or data is more stable inside the memory and as discus here in this work we used separate beltline to perform read and write operation to improve SNM of memory. The SNM is calculated using Origin software by using butterfly curve method. In the SRAM cell structure, the DC transfer characteristics of the two back to back in feedback connected CMOS inverters are used to make the butterfly curve as shown in Fig. 1.

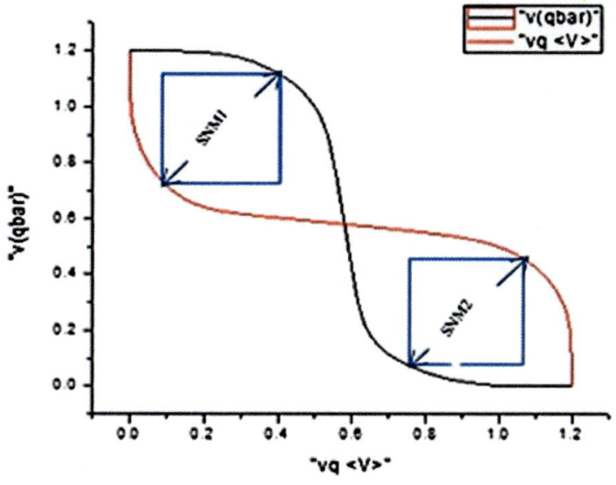

Fig. 1 Butterfly curve for calculating SNM of SRAM cell

During the data read operation, the applied maximum DC noise voltage at storage node Q and Qbar that an SRAM cell can tolerate without flipping of stored data is defined as the read SNM of the SRAM cell.

The write SNM is defined as, how fast the bitline voltage flip, the stored information at store node Q and Qbar.

Hold SNM is defined as maximum DC noise voltage at node Q and Qbar that an SRAM cell can tolerate without losing data during an idle state.

$$\text{SNM} = \min(\text{SNM1}, \text{SNM2}).$$

2.2 *Delay*

Here we used write and read delay calculation for the characterization of proposed memory cell during data read and write operation. The write delay of the memory cell is defined as time required for writing the date in memory [9]. For write "1", operation worldline is kept as high and bitline and bitline bar is kept as "1" and "0" respectively. For the read delay calculation precharged the bitline and bitline bar line with voltage "1" and read delay is defined as how fast the difference in bitlne and bilinebar voltage is sense by the sense amplifier, this is according to the stored information in the memory cell.

2.3 Leakage Current

Leakage current [10] is defined as the current that flows in a device, which is in "off" state, where no current will flow ideally.

Similarly, in SRAM cells during the write, read and hold operation leakage current flows through the OFF transistor(s), here we calculated the leakage current in SRAM cell that flows through the transistor, which is in "off" state during write, read and hold operations.

2.4 Static Power Dissipation

The write and read power dissipations [11, 12] are defined as the power consumed by the SRAM cell during its data write and read operations.

The write and read power dissipations are calculated using the leakage current during write and read operations respectively.

$$P_{static} = V_{dd} \times I_{leakage}$$

3 Asymmetrical Ground-Gated 7T SRAM Cell Characterization

An asymmetrical ground-gated 7T SRAM cell (Asym7T) is shown in Fig. 2, to achieve high stability and low leakage current, SLEEP mode of operation is proposed here. To improve the read stability, separate read port with control signal RWL is used. The read port consists of N4 and N5, which are read bitline access transistor and read transistor respectively. To reduce the leakage current of read bitline, a high threshold voltage (HVT) transistor, Nsl connected to the source of the read transistor (N4) and controlled by control signal SLEEP1. A write-assist circuitry that consists of Pch and Nwr is connected to the source of N1 and controlled by the control signal WWLB. The sleep transistor is turned ON in SLEEP mode (No read and write operation) and due to stacking of N4 and Ns1 helps in reducing the leakage current.

During the read "0" operation through the memory cell, the node voltage at Qb is "1", due to high node voltage at Qb the transistor N4 is turn ON. As for the read operation, RWL signal is high, the transistor N5 is also in ON state and read bitline start discharged through the N5, N4 and Sleep transistor. The RBL voltage is forced into a single-ended sense amplifier [13], which is shown in Fig. 3 to read the data stored in the node Q of the SRAM cell.

The write "0" and write "1" operations are performed as same as in case of conventional 6T SRAM Cell. Here for the write operation, WWLB signal is high

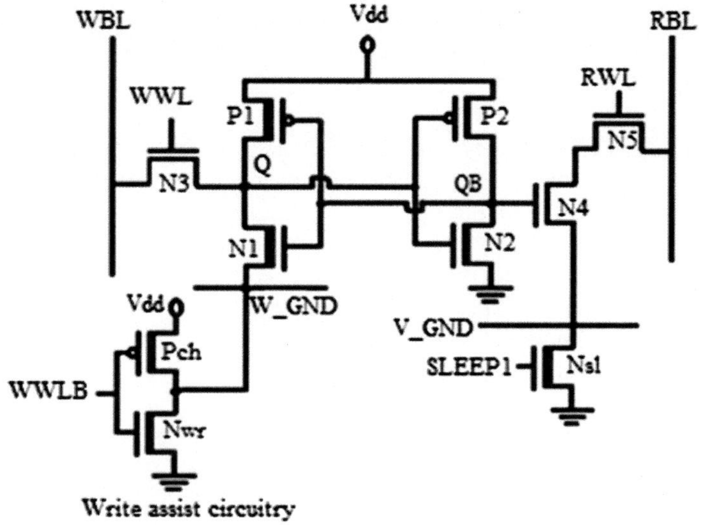

Fig. 2 Schematic of Asym7T SRAM cell. *HVT transistors are shown with thick channel region

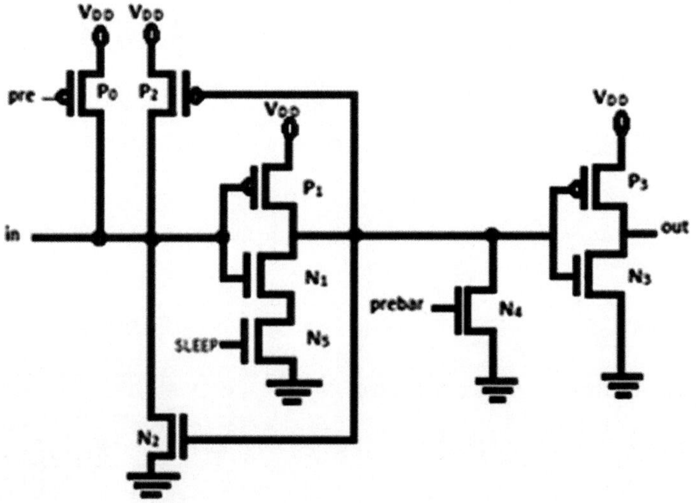

Fig. 3 Single-ended sense amplifier

that is the input to the inverter design by Pch and Nwr PMOS and NMOS respectively. Because of high input signal during write operation, the inverter output signal is Low that may be consider as ground (Low Voltage) for write operation.

The single-ended sense amplifier shown in Fig. 3 is composed of two data nodes "in" and "out" and three control nodes "pre", "prebar" and "SLEEP". The data nodes are input and output to the sense amplifier and its working is as follows: "pre" equalizes the data nodes. The SRAM cell that is being read is asserted and node "in" experiences a small voltage drop. At the beginning of each cycle, "in" is precharged to Vdd. "pre" is kept high for read operation. If "0" is stored in Qbar node of the SRAM cell, the voltage on RBL node is maintained at Vdd. "Out" of the sense amplifier has the same voltage as the RBL voltage and detects the voltage drop in it. P2 is used to prevent the data loss when "out" is at high voltage (Vdd). The voltage on the RBL is transferred to the input of sense amplifier "in".

The layout design and RCX extracted view of Asym7T SRAM cell are shown in Fig. 4a and b respectively. RCX extracted view shows the parasitic capacitance and resistance due to metals and polyline interconnects.

The post layout timing diagram of the Asym7T SRAM cell is shown in Fig. 5a. Table 1 lists the control signal of Asym7T SRAM cell for different operation conditions.

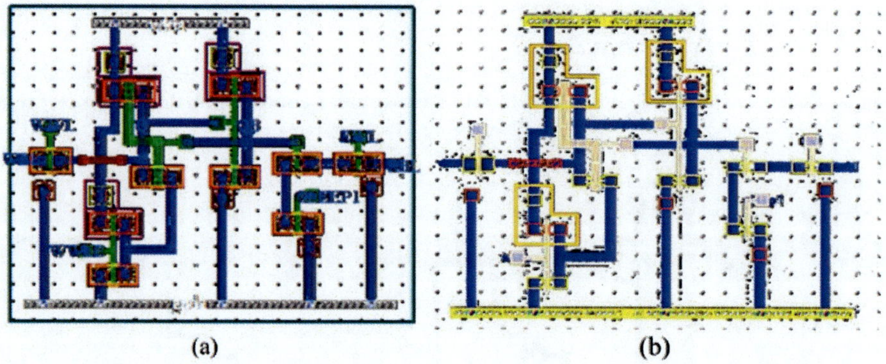

Fig. 4 **a** Layout of Asym7T SRAM cell. **b** RCX extracted view of Asym7T SRAM cell

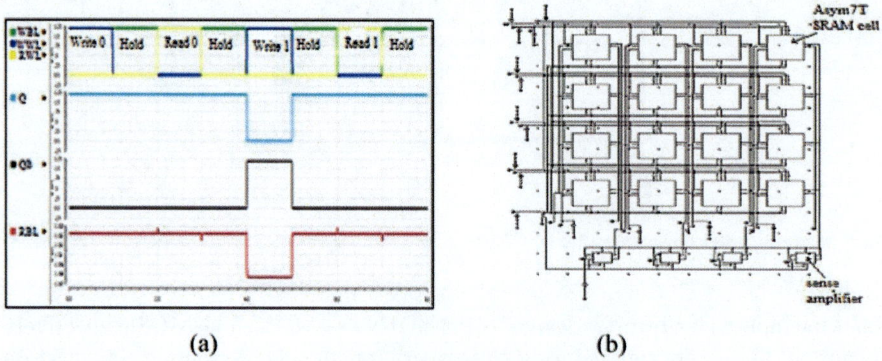

Fig. 5 **a** Post layout timing diagram of Asym7T SRAM cell. **b** Design of memory architecture of 4 × 4 memory array

Table 1 Control signals for operation of Asym7T SRAM cell

Signals	Hold	Write "1"	Write "0"	Read "1"	Read "0"
RWL	GND	GND	GND	V_{dd}	V_{dd}
WBL	V_{dd}	V_{dd}	GND	V_{dd}	GND
RBL	GND	V_{dd}	GND	V_{dd}	V_{dd}
WWL	GND	V_{dd}	V_{dd}	GND	GND
WWLB	V_{dd}	GND	GND	V_{dd}	V_{dd}

Table 2 Process corner simulation results of Asym7T SRAM cell

Parameters	SS	SF	FS	FF
Delay (nS)	37.4	25.7	27.7	14.4
Leakage current (pA)	1.23	1.54	1.24	1.26
Power dissipation (pW)	1.47	1.84	1.48	1.51

Table 3 Pre and post layout simulation results of Asym7T SRAM cell

Parameters	Pre layout	Post layout
Write delay (nS)	56.33	112.1
Write leakage current (pA)	1.24	10.6
Write power (pW)	1.48	12
Write SNM (mV)	608.02	600
Read delay (nS)	6.6	6.88
Read leakage current (pA)	0.5	0.51
Read power (pW)	0.6	0.612
Read SNM (mV)	466.6	452.4
Hold SNM (mV)	763.56	749.5

The circuit simulation of the proposed Asym7T SRAM cell is performed at various process corners is shown in Table 2. Where SS, SF, FS and FF denote slow–slow, slow–fast, fast–slow and fast–fast corner simulation.

Pre and post layout simulation results after RC extraction are also presented in Table 3.

4 Memory Array and Architecture

In this section, 4 × 4 memory array and memory array with all peripheral circuitry using Asym7T SRAM cell is presented. The memory array has separate output ports, wordline and bitline for each column.

To read the data from the array, single-ended sense amplifiers are used for each column of the memory array. A 4 × 4 memory array architecture using Asym7T SRAM cell is shown in Fig. 5b. For write operation, simulation waveform result of 4 × 4 memory array architecture is shown in Fig. 6a. Pre and post layout simulation results of 4 × 4 Memory array as write delay, read delay, write leakage current and read leakage current are shown in Table 4.

The memory architecture of 4 × 4 memory array consists of a row decoder to select the wordlines of the memory array. Single wordline is selected at any given time according to address provided by address decoder. Single-ended sense amplifiers are connected with each column of the memory array to read the data stored in the cells. Memory architecture along with peripheral circuitry using Asym7T SRAM cell is shown in Fig. 6b.

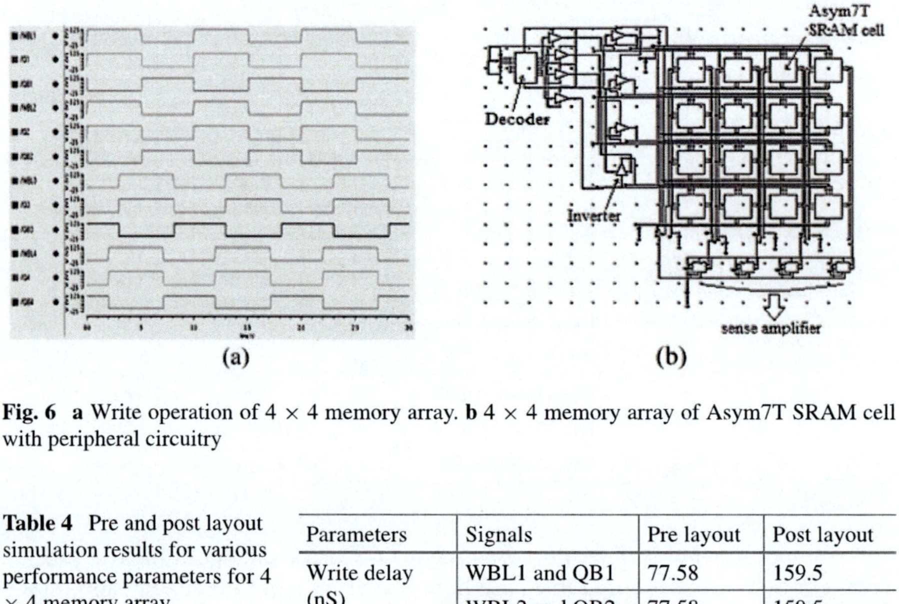

Fig. 6 **a** Write operation of 4 × 4 memory array. **b** 4 × 4 memory array of Asym7T SRAM cell with peripheral circuitry

Table 4 Pre and post layout simulation results for various performance parameters for 4 × 4 memory array

Parameters	Signals	Pre layout	Post layout
Write delay (nS)	WBL1 and QB1	77.58	159.5
	WBL2 and QB2	77.58	159.5
	WBL3 and QB3	76.76	154
	WBL4 and QB4	78.77	159.7
Read delay (mS)	Q1 and Out1	1.625	1.626
	Q2 and Out2	1.62	1.63
	Q3 and Out3	1.62	1.63
	Q4 and Out4	1.625	1.626
Write leakage current (fA)	–	0.117	0.139
Read leakage current (pA)	–	0.48	0.481

Table 5 Comparative study of Asym7T with ground-gated 6T SRAM cell for pre layout simulation results

Parameters	Ground-gated 6T [12]	Asym7T
Write delay (nS)	18.6	56.33
Write leakage current (nA)	40.5	0.001
Write power (nW)	48.6	0.0015
Write SNM (mV)	197.9	608.02
Read delay (nS)	0.68	6.6
Read leakage current (pA)	0.16	0.5
Read power (pW)	0.196	0.6
Read SNM (mV)	223.1	466.6
Hold SNM (mV)	408	763.56

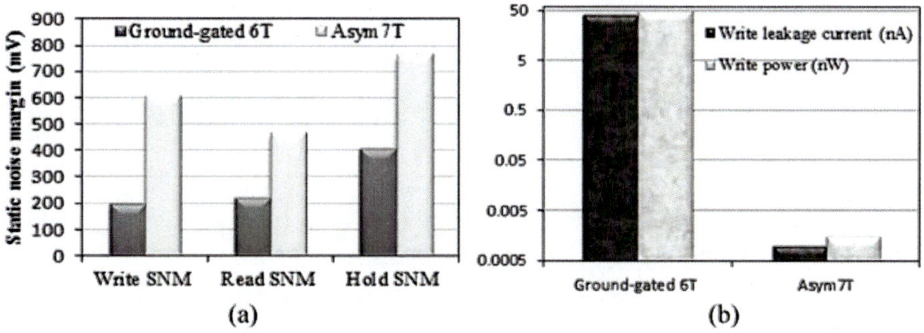

Fig. 7 **a** Static noise margins of ground-gated 6T and Asym7T SAM cells. **b** Write leakage current and power of ground-gated 6T and Asym7T SAM cells

WBL1, WBL2, WBL3 and WBL4 are the write bitlines of each column of array. Q1, Q2, Q3, Q4, QB1, QB2, QB3 and QB4 are the output nodes of the SRAM cells in the array. Out1, Out2, Out3 and Out4 are the output nodes of sense amplifiers connected to the columns of 4 × 4 memory array. A comparative study of Asym7T SRAM cell with ground-gated 6T SRAM [12] cell is shown in Table 5, here we considered all HVT transistors for pre layout simulation. The static noise margins of ground-gated 6T and proposed Asym7T SRAM cell are shown in Fig. 7a, write leakage current and write power are shown in Fig. 7b.

5 Conclusion

In this paper, the application of MTCMOS technique is used in Asym7T SRAM cell. The leakage current and power dissipation get reduced and stability gets enhanced by using MTCMOS technique. All the transistors are taken as HVT for the best results for all the characteristics for memory design. Process corner simulation is done in which FF simulation has the least delay i.e. 14.4 ns. Pre and post layout simulation of Asym7T SRAM cell is done for better memory design. Post layout simulation results are degraded as compared to pre layout results.

Asym7T SRAM cell has very less leakage current as compared to ground-gated 6T SRAM cell.

Asym7T SRAM cell has higher write, read and hold SNM up to 67.45, 52.18 and 46.56% as compared to ground-gated 6T SRAM cell shows that the proposed Asym7T SRAM cell has high stability.

A 4 × 4 memory array and complete memory architecture are designed using Asym7T SRAM cell. All required peripheral circuit for memory array design is designed using Cadence EDA tool and pre and post layout simulation results are presented in the paper.

Acknowledgements This work was supported by, Teachers Associateship for Research Excellence (TARE) scheme of Science and Engineering Research Board (SERB) DST, Government of India, New Delhi. File No: SERB/F/11697/2018-19 dated 27 February 2019.

References

1. Pavlov, A., Sachdev, M.: CMOS SRAM Circuit Design and Parametric Test in Nano scaled Technologies: Process-aware SRAM Design and Test. Springer Science and Business Media (2008)
2. Mutoh, S., et al.: 1-V power supply high-speed digital circuit technology with multi-threshold voltage CMOS. IEEE J. Solid State Circuits, New York **30**(8), 847–854 (1995)
3. Wei, L., et al.: Design and optimization of dual-threshold circuits for low-voltage low-power applications. IEEE Trans. VLSI Syst. New York **7**(1), 16–24 (1999)
4. Seevinck, E., List, F. J., Lohstroh, J.: Statis-noise margin analysis of MOS SRAM cells. IEEE J. Solid-State Circuits **SC-22**(5) (1987)
5. Sanvale, P., Gupta, N., Neema, V., Shahc, A.P., Vishvakarma, S.K.: An improved read-assist energy efficient single ended P-P-N based 10T SRAM cell for wireless sensor network. Microelectron. J. **92**, 104611 (2019)
6. Parihar, P., Gupta, N., Neema, V., Singh, P.: Investigation of MTCMOS 6T SRAM cell for ultra-low power applications. In: 2nd International Conference on Nano-electronics, Circuits & Communication Systems (NCCS-2016), pp. 49, 25–26 (2016)
7. Jiao, H., Qiu, Y., Kursun, V.: Variability-aware 7T SRAM circuit with low leakage high data stability SLEEP mode. IEEE J. Solid-State Circuits **43** (2008)
8. Oniciuc, L., Andrei, P.: Random doping fluctuation effects on static noise margins of 6TSRAM cells. In: Solid-State and Integrated Circuit Technology, pp. 488–491 (2008)
9. (Steve) Kang, S.-M., Leblebigi, Y.: CMOS Digital Integrated Circuits: Analysis and Design, Ch. 10, pp. 198–200. McGraw-Hill Higher Education (2003)
10. Mukhopadhyay, S., Raychoudhury, A., Roy, K.: Accurate Estimation of total leakage current in scaled CMOS logic circuits based on compact current modelling. In: Proceedings of Design Automation Conference, Anaheim, California, USA, pp. 169–174 (2003)
11. Tawfik, S.A.: Digital circuit methodologies for low power and robust nanoscale integration. Ph.D. thesis, University of Wisconsin, Madison (2009)

12. Liu, D., Svenson, C.: Power consumption estimation in CMOS VLSI chips. IEEE J. Solid State Circuits **29**(6), 663–670 (1994)
13. Jiao, H., Qiu, Y., Kursan, V.: Low power and robust memory circuits with asymmetrical ground gating. Microelectron. J. **48**, pp. 109–119 (2016)

Data-Aware Near Subthreshold 10 T SRAM Cell for Ultra-Low Power Application

Priyanka Sharma and Vaibhav Neema

1 Introduction

As technology grows, the various applications of artificial intelligence and the Internet of things become more popular. To enable this technology for mobile-based applications or wireless sensor-based application, battery power is limiting its processing capability. To overcome this situation or to improve the computing power at the sensor node, the only solution is by the use of ultra-low power circuits and memory. If on chip memory will work at ultra-low power, then the power requirement of the sensor node will significantly reduce. The design and operation of memory in the subthreshold region are one of the possible solutions to get the low power dissipation during the read, write, and hold operations [1] in memory circuit. But at the lower supply voltage, data stability in memory is the biggest challenge for the designers. The Six Transistor (6 T) SRAM cell is shown in Fig. 1a and it is observed that it fails to operate at lower supply voltage [2]. However, the impact of process variation becomes more substantial for low supply voltage circuits and it becomes even exponential in the subthreshold region [3]. Threshold voltage variation in SRAM cell due to random dopant various during the fabrication process will increase cell failure rate [4]. Threshold voltage variation cause by increasing leakage current and there is inverse exponential relation between leakage current and threshold voltage [5]. For the reduction of leakage current bitline leakage minimizing techniques [6, 7] leakage reduction using sleep transistors [8, 9] and multi-threshold devices [10, 11] have been introduced. Therefore, to overcome design challenges for designing

P. Sharma · V. Neema (✉)
Electronics and Telecommunication Engineering Department, Institute of Engineering and Technology- Devi Ahilya Vishwavidyalaya, Indore 452017, India
e-mail: vneema@ietdavv.edu.in

P. Sharma
e-mail: psharma@ietdavv.edu.in

A. Biswas et al. (eds.), *Microelectronics, Circuits and Systems*, Lecture Notes in Electrical Engineering 755, https://doi.org/10.1007/978-981-16-1570-2_13

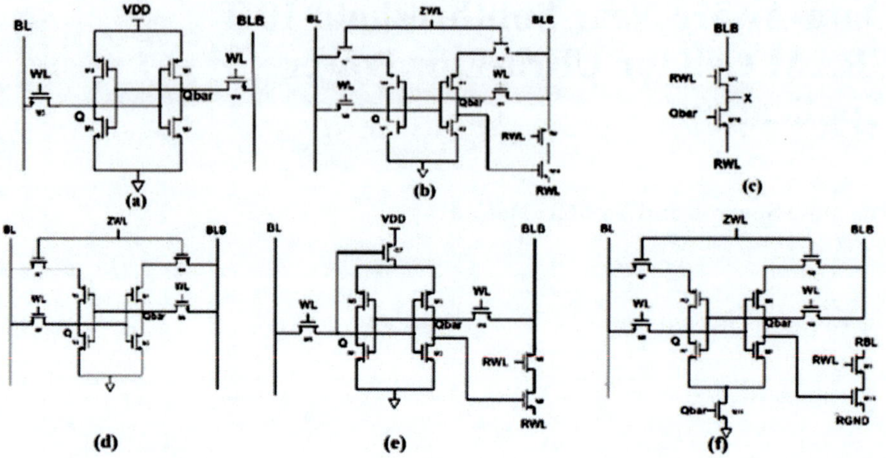

Fig. 1 Circuit diagram **a** Conventional Six Transistor (6 T) SRAM cell, **b** 10 T SRAM cell (proposed), **c** circuit for read buffer, **d** D2AP8T, **e** Vth_9T, **f** D2P11T

low power memory, various researchers proposed several SRAM cell architectures in recent years such as D2AP8T [12], Vth_9T [13], and D2P11T [14].

The given state-of-art SRAM cells, D2AP8T, Vth_9T, and D2P11T are differential reading and writing. The D2AP8T uses differential data-aware power-supplied (D2AP). This D2AP reduces power consumption at low VDD but reduces read stability due to common read and write word line. D2P11T SRAM cell has been proposed to increase the read stability, which uses the separate read buffer configuration. This read buffer consists of two NMOS stacked transistors that increase read stability and reduces power consumption in standby mode, but this configuration requires the number of control signals that increase complexity in generating that control signals and increasing area. Another configuration to decreasing power consumption at low is VDD Vth_9T. The Vth_9TSRAM cell has been proposed low threshold transistor in read path to increasing read access time and control leakage current during reading and holds operation to preventing standby mode power dissipation. The Write 1 Static Noise Margin (WSNM 1) of the cell degrades at low supply voltage [16]. Hence, for AI, IoT and other various applications require SRAM cells with low leakage and high reliability at lower supply voltage. The key highlights of the proposed work are as follows:

i. The use of data drove power supply leads to reduce the leakage power dissipation and improvement in the write noise margin of the proposed cell is calculated.
ii. During the write data operation in the memory cell, one of the inverters in cross-couple mode gets zero supply voltage that reduces the effort to flip the store data in memory.
iii. The use of read decouple and read buffer will increase the read static noise margin and read speed.

iv. The use of high threshold transistor in cross-couple inverter effectively reduces the leakage current during the hold operation.

The organization of the research paper is as follows. In Sect. 2, we elaborate the proposed 10 T SRAM cell. In Sect. 3, simulation results are presented. Section 4 concludes the manuscript.

2 Proposed 10 T SRAM Cell

The circuit diagram of the data-aware proposed 10 T SRAM cell is shown in Fig. 1b. The write, hold, and read operations in the proposed memory cell are performed using states of control signals are shown in Table 1. Here supply volatge V =300 mV is used for circuit to operate in low power mode, and the memory is also data driven. The praposed cell consist of two back to back connected inverters to store the bit information, which have high threshold transistors (M1, M2, M3, and M4). The purpose of using high Vth transistors is to minimize the leakage current during hold state. The transistors M9 and M10 are considered as low Vth transistors.

Here, M5 and M6 are access transistors controlled by control signal WL. Control signal WL is logic high during write operations and logic low during hold and read operations. M9 and M10 are read decouple transistors that are controlled by control signal RWL and Qbar. RWL control signal is logic low during the read operation and Qbar node voltage is data-dependent. The read static noise margin can be improved by the use of read decouple transistors. In conventional 6 T SRAM during the read operation, bitline or bitline_bar gets discharged through the access and pull-down transistor node voltage is data dependent. The read decouple circuit consist of transistor M9 and M10 hence to discharge bitline or bitline_bar and to improve the read static noise margin pull-down transistor must have high current driving strength. In the proposed memory cell read operation achieve higher read static noise margin without sizing of pull-down transistor. In conventional 6 T SRAM cell, write operation is performed by bit flipping at node Q and Qbar through the discharging of Q or Qbar node voltage. The bitline and bitline_bar are set to be logic low or high according to the information to be write in the memory cell. The write operation is guided through the pull-up transistor and access transistor and to get the high write noise margin, the access transistor must have high current driving strength. To achieve the write noise margin independent of the cell ratio or sizing of the transistor, here, in the proposed cell, the supply voltage VDD is provided through

Table 1 Control signals status of proposed 10 T cell

Operation	WL	ZWL	RWL
Read	0	0	0
Write	1	0	1
Hold	0	0	1

the M7 and M8 pMOS transistors controlled by the control signal ZWL and source of the transistor is connected with bitline and bitline_bar. The control signal ZWL is at logic low value during the read, write, and hold operations. According to the stored information in memory cell, transistors M7 and M8 provide the supply voltage for write operation. For write '1' operation, bitline voltage is charged at VDD bitline voltage discharged to gnd. Hence, for node Q, pull-up transistor M3 will get the voltage VDD and node Qbar to pull-up transistor M4 will get zero voltage. The zero supply voltage to the pull-up transistor M4 turns off the transistor, due to off mode of M4 transistor, data at the node Q will easily flip and desire information is written over the memory cell. No direct supply voltage is provided to the memory cell, due to this, leakage current is also suppressed, hence minimum leakage power dissipation is achieved.

2.1 Design and Analysis of Read Buffer

Read buffer circuit diagram is shown in Fig. 1c where two transistors M9 (PMOS) and M10 (NMOS) are used. Here both the read buffer transistors are conducting in subthreshold region due to lower supply voltage. Due to subthreshold mode of conduction, channel between source and drain is in week inversion state and drain current flows by diffusion only [12]. The subthreshold drain current flow through the transistors is presented by Eq. (1).

$$i_{\text{Sub}} = I_0 \exp\left(\frac{V_{GS} - V_{TH}}{nV_T}\right)\left(1 - \exp\left(\frac{-V_{DS}}{V_T}\right)\right) \tag{1}$$

where *I0* is the subthreshold current, which depends on the technology and is given as in Eq. (2):

$$I_0 = \mu c_{ox}\frac{W}{L}(n-1)V_T^2 \tag{2}$$

where n subthreshold swing factor, *VT* thermal voltage is given as KT/q, μ mobility, *WL* aspect ratio. Equation (3) shows the effective threshold voltage of any short channel device, which is the function of applied drain-source voltage *VDS* and body-source voltage *VBS*:

$$V_{TH} = V_{t0} - \lambda_{DS}V_{DS} - \lambda_{BS}V_{BS} \tag{3}$$

where in Eq. (3), *Vt0* is the threshold voltage of the device without considering the drain-induced barrier lowering (DIBL) and body effect, due to consideration of DIBL and body biasing effect λ*DS* > 0 and λ*BS* > 0 respectively.

M9 and M10 transistors are operating in subthreshold region, and both are connected in series hence equal drain current flows through both the transistors. The drain current flowing through the transistor M10 is calculated by applying KCL at node X of Fig. 1c. For read '0' operation, node voltages for M10 transistor are as follows:

$Qbar = V_{DD}, V_{RWL} = 0, V_{GS} = V_{DD} V_{DS} = V_x, V_{BS} = 0$

For finding the value of threshold voltage for M10 transistor during read '0', we will put the value of above parameters in Eq. (3) and get the threshold voltage of M10 during read '0' as shown in Eq. (4)

$$V_{\text{THread}} = V_{t0} - \lambda_{DS} V_x \tag{4}$$

By combining Eqs. 1, 2 and 4, I_{read} is the final read current of the proposed SRAM cell determine, which is the drain current of transistor M10 and calculated as:

$$I_{\text{read}} = \mu c_{ox} \left(\frac{W}{L}\right)_{10} (n-1) V_T^2 \exp\left[\frac{V_{DD} - V_{t0} + \lambda_{DS} V_x}{n V_T}\right] \cdot \left[1 - \exp\left(\frac{-V_x}{V_T}\right)\right] \tag{5}$$

During the hold '0' operation, the various node voltage values for the transistor M10 are calculated as: $Qbar = V_{DD}, V_{RWL} = V_{DD}, V_{GS} = 0, V_{DS} = V_x - V_{DD}, V_{BS} = -V_{DD}$

The threshold voltage for hold operation will be calculated by substituting values of the above parameters in Eq. (3) and we will get

$$V_{\text{THhold}} = V_{t0} - \lambda_{DS}(V_x - V_{DD}) + \lambda_{BS} V_{DD} \tag{6}$$

By combining Eqs. 1, 2, and 6, hold '0' current value through the proposed SRAM cell is presented in Eq. 7. I_{hold} is the drain current of transistor M10 is as follows

$$I_{\text{hold}} = \mu c_{ox} \left(\frac{W}{L}\right)_{10} (n-1) V_T^2 \exp\left[\frac{-V_{t0} + \lambda_{DS}(V_x - V_{DD}) - \lambda_{BS} V_{DD}}{n V_T}\right] \cdot \left[1 \exp\left(\frac{-V_x + V_{DD}}{V_T}\right)\right] \tag{7}$$

By comparing Eqs. (4) and (6), we found that $Vth, hold > V_{\text{th,read}}$ so, here we conclude that higher the threshold voltage minimizes the current that causes the low power dissipation during the hold operation and stability of storage node is increased due to proper sizing of the back to back connected inverter transistors. Low-threshold transistors M9 and M10 in the proposed design provide a low resistance path during read operation that causes an increment in current consequently reduces read access time.

2.2 Write Bit Operation of a Proposed SRAM Cell

1. Write ‘0’ bit operation: To write the bit ‘0’information in the proposed memory cell, the status of the control signal voltages is presented in Table 1. Initially, we assume that the cell stores the bit value ‘1’ and by writing ‘0’ we flip the node voltages at node Q and Qbar, here, initial voltages at the Q node at logic high and Qbar at a logic low level. By applying proper control signal voltages presented in Table 1, write ‘0’ operation initiates. During the write, operation read decouple circuit turnoff and for write ‘0’ operation, bitline is precharged at low and bitline bar is precharged with high voltage. As WL = 1 both access transistors are in ON state and bit flip initiate from the Q node that starts discharging through the access transistor and flips the store state by the inverter through positive feedback path.
2. Write ‘1’ bit operation: Bitline is precharged with high and bitline bar is precharged with low voltages for write ‘1’ bit operation. By applying control signal WL = 1, both the access transistors are turned on. The bit flip is initiated by the charging of node Q voltage through the access transistor. Once sufficient voltage is stored at the node Q, these flip the inverter state, and information ‘1’ is stored inside the SRAM cell.

In the proposed work during the write operation, supply voltage to the inverter cell is provided from the bitline and bitline_bar voltages and controls by the control circuit ZWL. During the write, ‘1’ operation, bitline voltage is precharged to high voltage hence inverter connected with bitline gets the high supply voltage and other inverters get the zero supply voltage. The advantage of zero supply voltage to the inverter is that the inverter is weak and the state can easily flip in it, which makes the write noise margin higher for the memory cell and also reduces the leakage current through the inverter.

2.3 Read Operation of a Proposed Cell

1. Read ‘0’ operation: The control signal voltages for Read ‘0’ operation of the proposed 10 T SRAM cell are presented in Table 1. During read operation, bitline and bitline_bar precharged to logic high voltage. As presented in Table 1, signal WL = ‘0’, both the access transistors M5 and M6 are in OFF state, and for read ‘0’ operation, node Q is at logic low and Qbar is at a logic high value. Due to Qbar at high voltage, transistor M10 of read decouple circuit is in turn ON condition means discharging of bitline_bar voltage, hence information ‘0’ is read.
2. Read ‘1’ operation: for the read ‘1’ operation, control signal voltages are the same as read ‘0’ operation. The advantage of the proposed cell observed during read operation, due to read decouple circuit data flip during the read operation is eliminated hence static read noise margin of the proposed cell is high.

2.4 *Hold Operation of a Proposed Cell*

In the hold operation of the proposed cell, the control single voltages are presented in Table 1 where WL, ZWL are logic low voltages, and RWL is logic high voltage. Due to logic low value of the control signal, WL cell is disconnected from the bitline and bitline_ bar lines because both the access transistors M5 and M6 are in turn OFF condition. The control signal RWL is at logic high value during the hold operation, turn OFF the M9 of the read decouple circuit that prevents any discharging of bitline bar voltage during hold operation. The control signal ZWL at logic low voltage provides the power supply to the cross-couple inverters through the transistors M7 and M8.

3 Simulation Results and Analysis

In this section, simulation results of various performance parameters of the memory cell are calculated and compared with the other state of art memory cell architecture as shown in Fig. 1. For the fair comparison, all the circuits are designed using 45 nm GPDK CMOS technology and simulated using Cadence design environment. The features comparison of these cells with the proposed 10 T is shown in Table 2.

Table 3 shows the simulated results and comparison of proposed 10 T with the conventional 6 T and other state-of-art SRAM cells (D2AP) 8 T [12], (Vth_9T) 9 T

Table 2 Comparison of SRAM cell features

Features	6 T	D2AP8 T	VTH_9 T	(D2P) 11 T	Proposed 10 T
Reading/writing	Diff./diff	Diff./diff	SE./diff	SE/diff	SE/diff
Control signal	WL	ZWL	WL, RWL	WL, RWL	WL
Bit lines	2 BL	2 BL	2 BL	2 BL, 1 RBL	2 BL

(BL) bit line, (Diff) Differential, (RWL) read word line (SE), single ended, (WL) word line, (RGND) read ground, (ZWL) control signal for bit lines powering

Table 3 Comparative analysis of various SRAM cell architectures

VDD = 0.3 mV, Temp. = 27°C@ 45 nm TN	Proposed 10 T	6 T	8 T	9 T	11 T
Leakage power (pW)	0.843	1.722	1.722	1.632	1.51
RSNM (mV)	101.66	Unstable	Unstable	77.52	38.75
HSNM (mV)	101.66	110.11	109.66	77.44	38.75
WSNM 0 (mV)	188.31	88.68	173.84	109	203
WSNM 1 (mV)	188.31	88.68	173.84	155.55	195.87
Read delay	94.96 (ps)	1.35 (ns)	1.85 (ns)	44.61 (ps)	44.61 (ps)

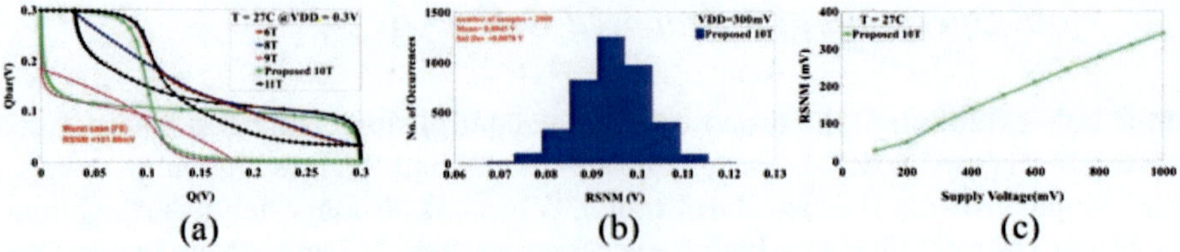

Fig. 2 **a** Comparison of read butterfly curves for various SRAM cells at worst process corner (FS) at 0.3 V. **b** RSNM Histogram of proposed cell at 0.3 V supply voltage with 2000 Monte Carlo simulation. **c** RSNM at different supply voltages

[13], and (D2P) 11 T [14]. Here we considered uniform device sizing at 300 mV supply voltage and operating temperature of 27 °C with worst corner analysis.

3.1 Read Stability

The maximum DC noise voltage that a SRAM cell can tolerate without flipping the stored data during the read operation is known as Read SNM. Here in the research work, static noise margin is calculated using the butterfly curve method where noise margin is defined as the largest square that can best fit into the smaller lobe of the butterfly curve [15]. Figure 2a shows the comparison of worst-case Fast Slow (FS) corner read butterfly curve at the supply voltage 300 mV. The Read SNM of the proposed 10 T memory cell is calculated as 101.66 mV, which has the highest value of Read SNM with compare to other state-of-art SRAM cells. The conventional 6 T and (D2AP) 8 T [12] are unstable at FS corner. The proposed memory cell shows the 1.31X and 2.62X improved in Read SNM as compared to 9 T Vth_9T [13] and (D2P) 11 T [14] techniques respectively. Here to show the stability of the obtained results, 2000 point Monte Carlo simulation at 300mv of supply voltage is performed and presented in Fig. 2b. The results obtained for Monte Carlo simulation represent that the mean is at 0.094 V with a standard deviation of 0.007 V. Figure 2c shows the variation on RSNM at different supply voltages.

3.2 Write Stability

Minimum bit line voltage required to flipping the state of the memory is defined as write static noise margin (WSNM). Here in the research work, we used butterfly curve method to determine the write SNM of the memory cell [15]. Figure 3a, b shows the comparative results of write butterfly curves for the proposed 10 T memory cell and other state-of-art SRAM cells at worst slow fast (SF) corner using 300 mV supply voltage for write '1' and write '0' operation. The 188.31 mV is the calculated value of write SNM for the proposed memory cell for write '0' and write '1'opeartions.

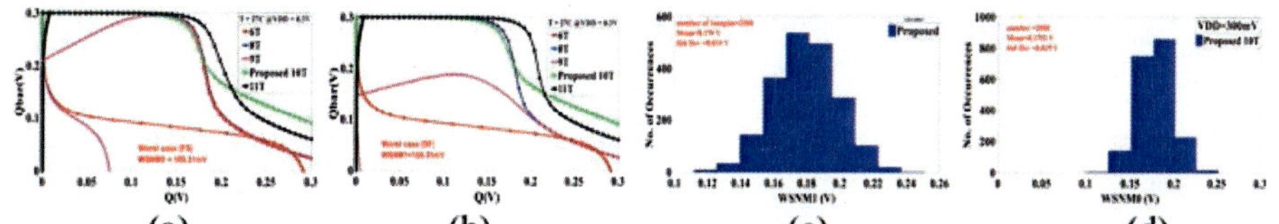

Fig. 3 Comparative results of **a** write '0' SNM, **b** write '1' SNM using butterfly curve method for various SRAM cells at worst process corner (SF) using 300 mV supply voltage. **c** Write '1' SNM **d** Write '0' SNM for 2000 Monte Carlo simulation for Write SNM of proposed cell using 300 mV supply voltage

For the process variation analysis on WSNM, 2000 point Monte Carlo simulation is performed and presented in Fig. 3c, d. From the simulation results, it is concluded that the proposed memory cell has a deviation of 0.019 V and mean of 0.179 V at 300 mV supply voltage for WSNM '1' and '0' operations.

3.3 *Hold Static Noise Margin*

The maximum DC voltage that SRAM cell can tolerate without losing data during the idle state is defined as Hold static noise margin (HSNM). The HSNM of the proposed memory cell is calculated as 101.90 mV for the FS corner at 300 mV supply voltage. We observed that in Fig. 4a, the HSNM is improved in proposed cell as compared to 9 T Vth_9T [13] and (D2P) 11 T [14] respectively. The proposed cell Hold SNM is 1.31X and 2.62X, which is 9% less from conventional 6 T SRAM.

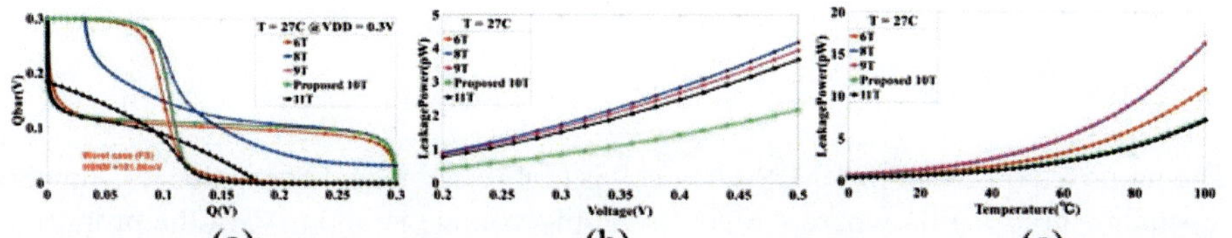

Fig. 4 **a** Comparative analysis of Hold operation using butterfly curve for various SRAM cells at worst process corner (FS) using 300 mV supply voltage. **b** Various SRAM cells, leakage power dissipation versus supply voltage variation. **c** Effect of temperature variation on leakage power dissipation for various SRAM cells

3.4 Leakage Power

During the data hold operation, power dissipated by the memory cell is defined as leakage power. The leakage power comparison of the SRAM cells at different supply voltages is presented in Fig. 4b. Due to the use of high threshold voltage transistor in cross-coupled inverter, data dependent power supply and read buffer circuit, the proposed cell drawn less leakage current and less power dissipate during hold operation. The leakage power is 0.84pW at 300 mV supply voltage and consume 50, 50, 48 and 46% less leakage power as compared with 6 T, (D2AP)8 T [12], Vth_9T [13] and (D2P) 11 T [14].

To check the thermal stability, the simulation of the proposed work is also performed for temperatures range from 0 °C to 100 °C at 300 mV supply voltage. As presented in Fig. 4c, from the obtained results, it is concluded that the proposed 10 T cell has equal leakage power as compared to (D2P) 11 T [14] and less leakage power as compared with rest of state of art cell at a higher temperature value.

3.5 Read Delay

The calculated read delay of various SRAM cells is shown in Table 3. The read delay of the proposed memory cell is lower than the 6 T and (D2AP)8 T memory cell but has slightly higher values as compared to the other 9 T and 10 T SRAM cell. The main cause of increase in delay for the proposed memory cell is because stacking of transistor during the read operation and due to use of high threshold transistor in read discharge path.

4 Conclusion

The proposed data-driven 10 T SRAM cell is highly stable and consumes less power in subthreshold region where the power supply voltage is 300 mV. In the proposed memory cell, the use of high threshold voltage transistors in cross-couple inverter reduces the leakage current of the cell and the use of read decouple circuit helps to improve the read stability of the memory cell. The feature of data-driven power supply in the proposed memory cell leads to an increase in write noise margin and also reduces the leakage current through the memory cell. The read noise margin and hold noise margin of the proposed cell at FS corner are calculated at 101.66 mV where conventional 6 T SRAM cell is unstable at 300 mV supply voltage. The write static noise margin of the proposed cell is calculated as 188.31 mV. The proposed memory cell shows the better noise handling capability at lower supply voltage during the read, write, and hold operations as compared to the other memory cell architectures.

Along with the high noise stability, the proposed memory cell has the lowest leakage current with a smaller area overhead.

Acknowledgements This research work is supported and funded under the TARE scheme of SERB, DST, New Delhi. File No: SERB/F/11697/2018-19 dated 27 February 2019,

References

1. Calhoun, B.H., Chandrakasan, A.P.: A 256-kb 65-nm sub-threshold SRAM design for ultra-low-voltage operation. IEEE J. Solid-State Circ. **42**(3), 680–688 (2007)
2. Bortolotti, D., Bartolini, A., Weis, C., Rossi, D. and Beninio, L.: Hybrid memory architecture for voltage scaling in ultra-low power multi-core biomedical processors. In 2014 Design, Automation & Test in Europe Conference & Exhibition (DATE) (pp. 1–6). IEEE (2014).
3. Wen, L., Li, Z., Li, Y.: Single-ended, robust 8T SRAM cell for low-voltage operation. Microelectron. J. **44**(8), 718–728 (2013)
4. Yamaoka, M., Maeda, N., Shinozaki, Y., Shimazaki, Y., Nii, K., Shimada, S. and Kawahara, T.: Low-power embedded SRAM modules with expanded margins for writing. In ISSCC, 2005 IEEE International Digest of Technical Papers. Solid-State Circuits Conference (pp. 480–611). IEEE (2005).
5. Wang, A., Calhoun, B.H., Chandrakasan, A.P: Sub-threshold design for ultra-low-power systems (vol. 95). Springer, New York (2006)
6. Verma, N., Chandrakasan, A.P.: A high-density 45 nm SRAM using small-signal non-strobed regenerative sensing. IEEE J. Solid-State Circ. **44**(1), 163–173 (2008)
7. Wu, J.J., Chen, Y.H., Chang, M.F., Chou, P.W., Chen, C.Y., Liao, H.J., Yamauchi, H.: A large σV_{TH}/VDD tolerant zigzag 8T SRAM with area-efficient decoupled differential sensing and fast write-back scheme. IEEE J. Solid-State Circ. **46**(4), 815–827 (2011)
8. Zhang, K., Bhattacharya, U., Chen, Z., Hamzaoglu, F., Murray, D., Vallepalli, N., Wang, Y., Zheng, B., Bohr, M.: SRAM design on 65nm CMOS technology with integrated leakage reduction scheme. In 2004 Symposium on VLSI Circuits. Digest of Technical Papers (IEEE Cat. No. 04CH37525) (pp. 294–295). IEEE (2004)
9. Wang, Y., Ahn, H.J., Bhattacharya, U., Chen, Z., Coan, T., Hamzaoglu, F., Hafez, W.M. et al: A 1.1 GHz 12σA/Mb-leakage SRAM Design in 65 nm ultra-low-power CMOS technology with integrated leakage reduction for mobile applications. IEEE J. Solid-State Circ. **43**(1), 172–179 (2008)
10. Liu, P., Wang, J., Phan, M., Garg, M., Zhang, R., Cassier, A., Chua-Eoan, L. et al.: A dual core oxide 8T SRAM cell with low Vccmin and dual voltage supplies in 45nm triple gate oxide and multi Vt CMOS for very high performance yet low leakage mobile SoC applications. In 2010 Symposium on VLSI Technology (pp. 135–136). IEEE (2010)
11. Diaz, C.H., Young, K.L., Hsu, J.H., Lin, J.C.H., Hou, C.S., Lin, C.T., Liawet, J.J. et al.: A 0.18/spl mu/m CMOS logic technology with dual gate oxide and low-k interconnect for high-performance and low-power applications. In 1999 Symposium on VLSI Technology. Digest of Technical Papers (IEEE Cat. No. 99CH36325) (pp. 11–12). IEEE (1999)

12. Chang, M.-F., Jui-Jen, Wu., Chen, K.-T., Chen, Y.-C., Chen, Y.-H., Lee, R., Liao, H.-J., Yamauchi, H.: A differential data-aware power-supplied (D^2AP) 8T SRAM cell with expanded write/read stabilities for lower VDDmin applications. IEEE J. Solid-State Circ. **45**(6), 1234–1245 (2010)
13. Moghaddam, M., Timarchi, S., Moaiyeri, M.H., Eshghi, M.: An ultra-low-power 9T SRAM cell based on threshold voltage techniques. Circ. Syst. Signal Process. **35**(5), 1437–1455 (2016)
14. Sharma, V., Gopal, M., Singh, P., Vishvakarma, S.K., Chouhan, S.S.: A robust, ultra low-power, data-dependent-power-supplied 11T SRAM cell with expanded read/write stabilities for internet-of-things applications. Analog Integr. Circ. Sig. Process **98**(2), 331–346 (2019)
15. Singh, P., Vishvakarma, S.K.: Ultra-low power, process-tolerant 10T (PT10T) SRAM with improved read/write ability for internet of things (IoT) applications. J. Low Power Electron. Appl. **7**(3), 24 (2017)
16. Sanvale, P., Gupta, N., Neema, V., Shah, A.P., Vishvakarma, S.K.: An improved read-assist energy efficient single ended PPN based 10T SRAM cell for wireless sensor network. Microelectron. J. **92,** 104611 (2019)
17. Pasandi, G., Fakhraie, S.M.: A 256-kb 9T near-threshold SRAM with 1k cells perbitlineand enhanced write and read operations. IEEE Trans. Very Large Scale Integr. (VLSI) Syst. **23**(11), 2438–2446 (2015).

Radiation Immune SRAM Cell for Deep Space Applications

Sparsh Koushik, Pawan Kumar Sahu, Shashank Kumar Dubey, and Aminul Islam

1 Introduction

There exist a large number of highly energized particles in the space environment. Unexpected errors or even failure of the complete system may come about if these highly excited particles hit the integrated circuits. The Static Random Access Memory, abbreviated as SRAM, carries with it a substantial amount of data, which unfortunately are extremely vulnerable to the radiations. Since the SRAM is a very crucial part of the aerospace electronics system, the study of radiation hardening and its impact on these memory cells become very important.

In deep submicron technologies, virtue of aggressive dimension scaling in the MOSFETs, SRAM cells are more prone to radiation-induced single event upsets due to large sensitive volume and low storage node capacitances. Earlier, the error-correcting codes are used to limit the soft error in SRAM. However, due to longer processing time and larger area overhead, error-correcting codes protected SRAM are not appropriate for aerospace applications. Due to smaller silicon area, lower power dissipation and shorter access delay are better options to enhance the single event upset (SEU) immunity as compared to traditionally used error-correcting codes. This motivates to design a radiation [1] hardened SRAM cell to provide robust operation even under process, voltage and temperature (PVT) variations and severe radiation environment in space with the help of double exponential current source. SRAM [2] cells occupy more than half of the chip area of modern-day high-performance ICs. As transistor dimension and spacing between the transistors is reduced due to CMOS technology scaling, multiple transistors are susceptible to the charge deposited from a single particle causing single event multiple node upsets [3]. This encourages

S. Koushik · P. K. Sahu · S. K. Dubey (✉) · A. Islam
Department of Electronics and Communication Engineering, Birla Institute of Technology, Mesra, Ranchi, Jharkhand, India

A. Islam
e-mail: aminulislam@bitmesra.ac.in

A. Biswas et al. (eds.), *Microelectronics, Circuits and Systems*, Lecture Notes in Electrical Engineering 755, https://doi.org/10.1007/978-981-16-1570-2_14

the need for soft-error resilient SRAM cells that can function even under extreme radiation conditions.

A variety of structures have been proposed that address the problem of radiation-induced soft errors. Authors in [4] have proposed two radiation-hardened-by-design (RHBD) memory cells (PS-10T and NS-10T) by using a stacked structure. However, these memory cells cannot provide fully single event upset (SEU) protection. NS-10T cell can only recover 0 → 1 SEU, whereas PS-10T cell has the capability of tolerating only 1 → 0 SEU. In [5], the authors proposed a differential read cell. 12T Dual Interlocked storage Cell (DICE-12T) was proposed in [6], which uses dual node feedback control to make it fully immune against SEU on a sensitive node. However, it is unable to recover single-event–multiple-node upsets (SEMNUs), which is becoming more critical reliability issue for emerging nanometer CMOS technology. The 12-transistor radiation hardened by design (RHD-12T) memory cell was proposed in [7]. Besides providing SEU immunity on any of its internal single nodes, it can also provide the SEMNUs immunity to some extent. Another RHBD-10T cell is recently proposed in [8] to provide area-efficient solution to SEU immunity. However, its radiation hardness performance in terms of SEMNUs is yet to be investigated. The previous solutions either show larger area overhead or limited SEU immunity. Therefore, they may not be suitable for aerospace applications where both area-efficient and highly reliable performances are required. Hence, radiation hardness by design is preferred for enhanced performance.

The remaining part of the paper has been organized as follows. Section 2 explains the proposed WARH12T memory cell structure and behavior, SEU recovery analysis, MEMNU recovery analysis. Section 3 explains the evaluation of the different parameters of the proposed cells and comparison cells. This includes access time evaluation and comparisons, cost comparison, SEU robustness verification and comparison, stability comparison (RSNM, WSNM and HSNM), read power comparison, write power comparison and hold power comparison. Section 4 contains the conclusion.

2 Proposed WARH12T Memory Cell

2.1 Cell Structure and Behavior

Figure 1 shows the schematic diagram of the proposed WARH12T SRAM cell. The proposed cell has four storage nodes Q, QN, VL, VR, four PMOS (P1, P2, P3, P4) and eight NMOS (N1, N2, N3, N4, N5, N6, N7, N8) to form a latch. Q and QN nodes are the storage nodes, VL and VR are the intermediate storage nodes. The transistors N5 and N6 are the access transistors, which are controlled by row-based WL signal and used for read and write operations. The transistors N7 and N8 are also the access transistors, which are connected with the intermediate node VR and VL, also connected with the bit lines BL and BLB, respectively, for easy write operations. The four transistors N5, N6, N7 and N8 improve the write capability of

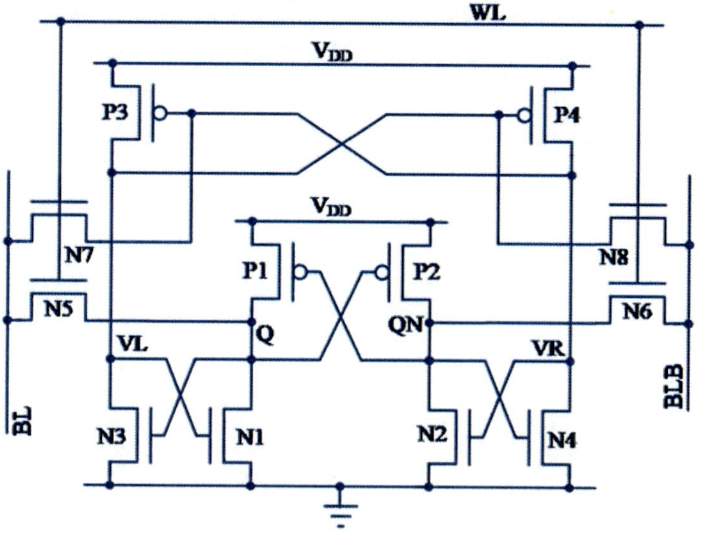

Fig. 1 Proposed WARH12T memory cell

the proposed cell. In order to achieve balance read and write operation, the cell ratio $(WN_1/LN_1)/(WN_5/LN_5)$ or $(WN_2/LN_2)/(WN_6/LN_6)$ is set to be 2.4 for better read stability. The pull-up ratio $(WP_1/LP_1)/(WN_5/LN_5)$ or $(WP_2/LP_2)/(WN_6/LN_6)$ is set to be 1.2. The second pull-up ratio $(WP_3/LP_3)/(WN_3/LN_3)$ or $(WP_4/LP_4)/(WN_4/LN_4)$ is set to be 4.5 for providing the better feedback and easy recovery of the storage nodes Q and QN. The NMOS transistors N3 and N1 or N2 and N4 are cross-coupled to pass the output storage node value to the feedback path, the PMOS transistor P3 and P4 are also cross-coupled and it is used as a positive feedback, which helps to recover the storage node value at Q and QN.

2.2 *SEU Recovery Analysis*

Taking the case when the state of the sensitive nodes Q, QN, VL and VR is 0, 1, 1 and 0, respectively.

(1) When the output storage node Q is affected by an SEU, and its potential changes from 0 to 1. Transistor P2 is OFF and transition N3 is ON momentarily, as P2 is OFF QN is at high impedance state and keeps its state as logic 1. Meanwhile N3 is ON and P3 is already ON making VL node unstable for a while and its potential is decided by the strength of P3 and N3 as we kept the strength of P3 is more than the strength of N3. So, the value of node VL becomes 1 and it will switch ON the transistor N1, which will help to recover the value at node Q immediately.

(2) When the output storage node QN is affected by an SEU and its potential changes from 1 to 0, transistor N4 is OFF and transistor P1 is ON temporarily. As N4 is OFF, node VR is at high impedance state and keeps its value as logic 0. Meanwhile P1 is ON and N1 is already ON, which will make node Q unstable for a while and its potential is decided by the strength of transistor P1 and N1, As we kept the strength of N1 is higher than P1, so the logic is decided by the pull-down transistor N1 and value at node Q is 0. It will switch ON the transistor P2 again and the node voltage of QN is recovered back to logic 1.

2.3 *MEMNU Recovery Analysis*

Multiple events $0 \rightarrow 1$ and $1 \rightarrow 0$ can occur at multiple nodes. Nodes Q and QN are capable of handling both 0 to 1 and 1 to 0 multiple events simultaneously. Taking the case when the state of all the nodes Q, QN, VL and VR is 0, 1, 1 and 0, respectively. When the Node Q and QN change from $0 \rightarrow 1$ and $1 \rightarrow 0$, respectively, transistor P2 is OFF and transistor N3 is ON momentarily, transistor N4 is OFF and transistor P1 is ON. As transistor N4 is OFF, VR is at high impedance state and keeps its state as logic 0. Meanwhile N3 is ON and P3 is already ON making VL node unstable and its potential is decided by the strength of P3 and N3, as we kept the strength of P3 more than the strength of N3. So the value at node VL becomes 1 and it will again switch ON the transistor N1 and try to pull down the potential at Q to zero but P1 is also ON due to upset at QN from $1 \rightarrow 0$, so the potential at node Q is now decided by the strength of P1 and N1 and as we kept the strength of N1 more than the strength of P1, node value of Q becomes 0 and it will switch ON the PMOS transistor P2, which will help to recover the value at node QN. In this way, with the help of feedback transistors P3 and P4, the node voltage of Q and QN is recovered very fast. This proposed WARH12T is capable enough to handle simultaneous upset at both the output-sensitive nodes Q and QN.

2.4 *Implementation*

The proposed WARH12T memory cell is implemented in 22 nm CMOS technology. The operating voltage for simulation is used as 1.2 V. The layout of the proposed cell WARH12T as well as the comparison cells is implemented in Microwind Version 3.1. All the measurement simulations of proposed cell and comparison cells are performed using SPICE. Double exponential current is also modeled in the SPICE, which mimics the same upsets as occurred in the SPACE. The layout of the proposed cell is shown in Fig. 2.

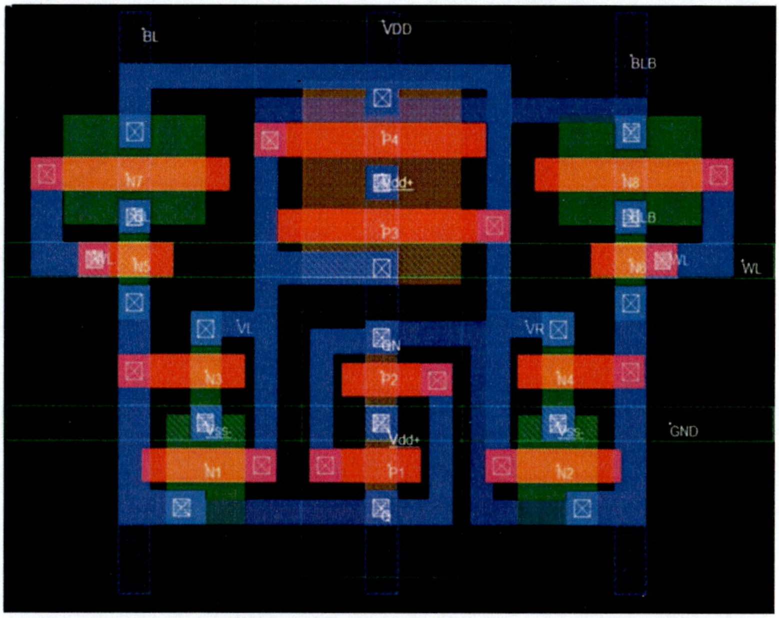

Fig. 2 Layout of the proposed WARH10T memory cell

3 Simulation Results and Discussion

3.1 Access Time Comparison

The read access time (TRA) is measured as the time taken for the voltage of BLB line to fall by 50 mV in the read condition, this time measurement starts from the time when WL starts to be asserted. Write access time is measured as the time taken for the voltage of a given storage node of the given SRAM to reach 90% of the V_{DD} when a write is attempted for that node, this time measurement starts from the time when the word line selected for the given SRAM starts to be asserted.

The TRA and TWA for the proposed structure are calculated at the supply voltages of 1.08, 1.14, 1.2, 1.26 and 1.32 V. The supply voltage of design is 1.2 V. The other values of supply voltage used are to account for supply voltage variation by 5 and 10%. Similarly, TRA and TWA have been calculated for all the comparison structures namely 6T, PS-10T, NS-10T, RHBD10T, RHD-12T and UTSC-12T at these five voltage values. The values of TRA and TWA of all these structures at different supply voltages can be visualized in Fig. 3a and b, respectively, to gain some insights.

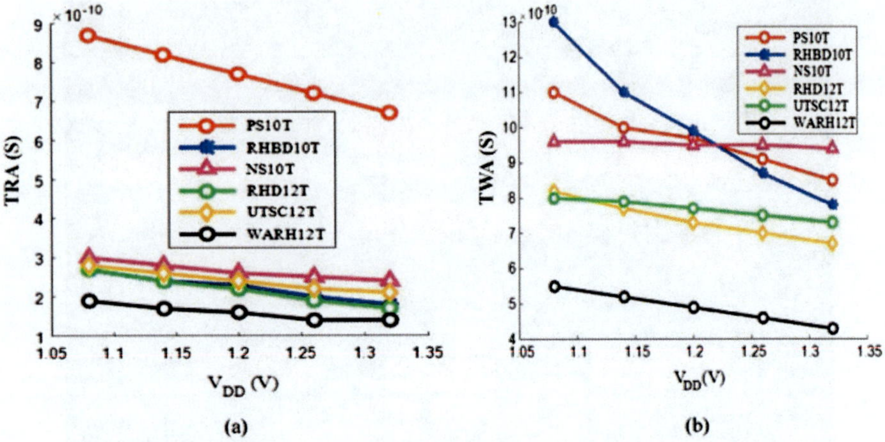

Fig. 3 Plot of **a** TRA, **b** TWA of proposed and comparison cells with V_{DD}

3.2 SEU Robustness Comparison

The SEU robustness was verified by using a double exponential current source. A double exponential current source was injected at the sensitive nodes. This was done to simulate the radiation strike or single event upset at these nodes. These sensitive nodes are characterized as the nodes, which are reverse biased [1].

To check the relative stability of the different structures to single event, we checked the maximum current, the circuit can tolerate before the stored data at the storage nodes of the SRAM cell changes or flips. The maximum value of the peak current for which the proposed WARH12T cell operated correctly is 870 uA. This maximum current tolerance for different comparison cells was found to be 91.63% smaller for PS-10T, 87.35% smaller for RHBD-10T, 90.52% smaller for NS-10T, 48.39% smaller for RHD-12T and 91.82% smaller for UTSC–12T as compared to the proposed cell at the proposed supply voltage of 1.2 V. It can be noticed that the radiation handling capability of the proposed WARH12T cell is significantly larger than the other comparison cells due to strong cross-coupled PMOS (P3 and P4) feedback, which does not allow to change the voltage at nodes VL and VR even if the output storage nodes Q and QN get affected by SEU, SEMNU and MEMNU.

3.3 Stability Comparison

Static noise margins for read, write and hold states have been used to quantify the stability of the proposed cell and hence for the comparison of the stability of the proposed cell with that of the existing or comparison cells. The static margins calculated are Read Static Noise Margin (RSNM), Write Static Noise Margin (WSNM)

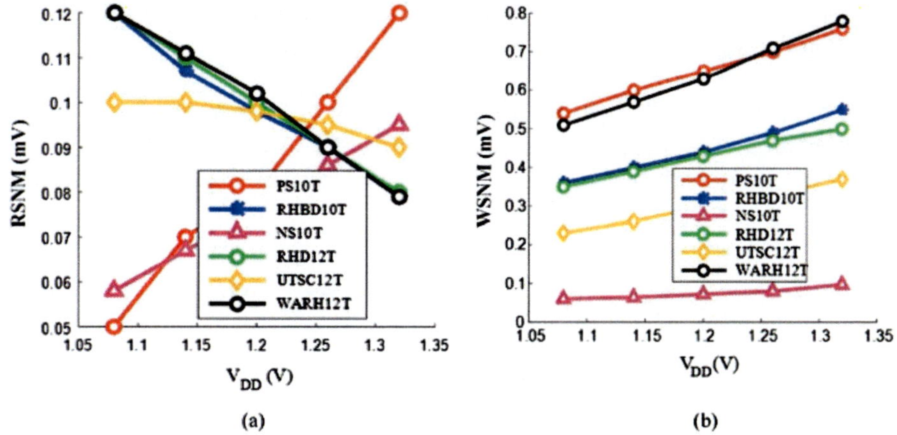

Fig. 4 Plot of **a** RSNM and **b** WSNM of proposed and comparison cells with V_{DD}

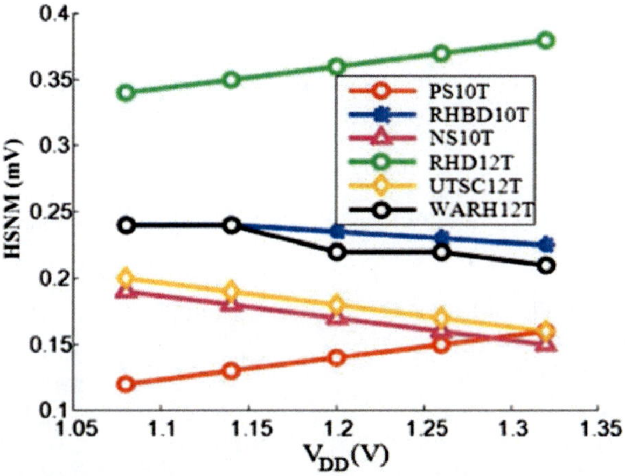

Fig. 5 Plot of HSNM of proposed and comparison cells with V_{DD}

and Hold Static Noise Margin (HSNM). The RSNM, WSNM and HSNM were calculated for the proposed and comparison cells at supply voltages of 1.08, 1.14, 1.20, 1.26 and 1.32 to study the behavior of the stability of the cells with voltage variation. The variation of the RSNM, WSNM and HSNM with voltage has been plotted in Figs. 4a, b and 5, respectively.

3.4 *Power Comparison*

For comparing the power consumption of the proposed cell WARH12T with that of existing cells, the power consumed by the proposed cell and the comparison cells was calculated during a read operation, write operation and hold state. These powers are termed as read power (RPWR), write power (WPWR) and hold power (HPWR). The read power quantifies the power dissipated during a read operation similarly the WPWR and HPWR represent the power dissipated during a write and hold operation. The values obtained for the RPWR, WPWR and HPWR for the proposed and comparison cells for different supply voltages are shown in Figs. 6a, b and 7, respectively.

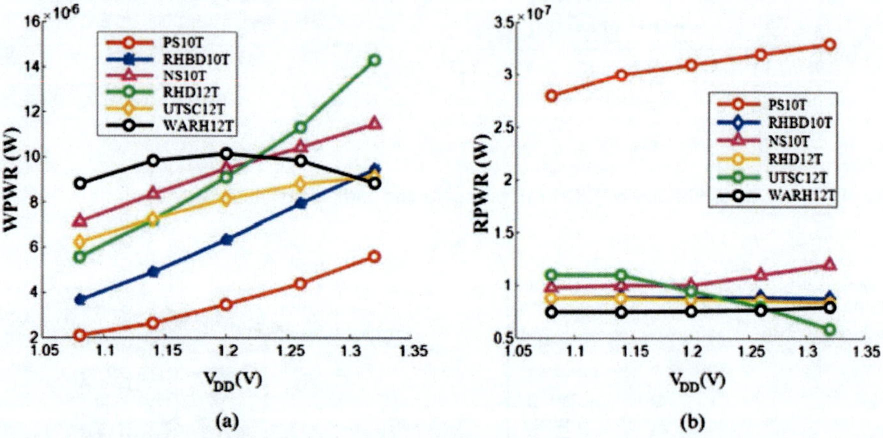

Fig. 6 Plot of **a** WPWR and **b** RPWR of proposed and comparison cells with V_{DD}

Fig. 7 Plot of HPWR of proposed and comparison cells with V_{DD}

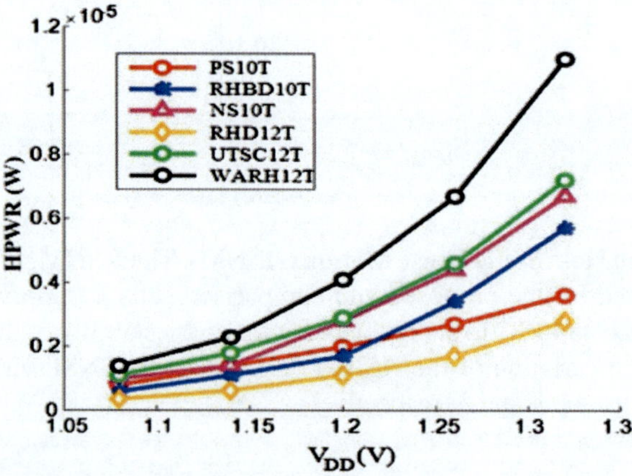

3.5 *TRA and TWA Variability*

This section talks about the standard deviation of read access time and write access time. This standard deviation is calculated using Monte Carlo simulations. The Monte Carlo simulations are run for 5000 iterations. This parameter gives an idea of how the read access time and write access time of the circuit deviate under PVT variations. The values of the standard deviation for TRA and TWA of the proposed and comparison cells at different supply voltages are represented in Figs. 8 and 9.

The standard deviation for the TRA of the proposed cell WARH12T at 1.2 V supply voltage is calculated to be 52 ps. From Fig. 8, it can be observed that the TRA variability of the proposed cell is similar in value to other comparison cells, although its value is slightly greater than that of the best performing cells. The standard deviation for the TWA of the proposed cell WARH12T at the proposed supply voltage of 1.2 V was found to be 51 ps. From Fig. 9, it can be concluded that

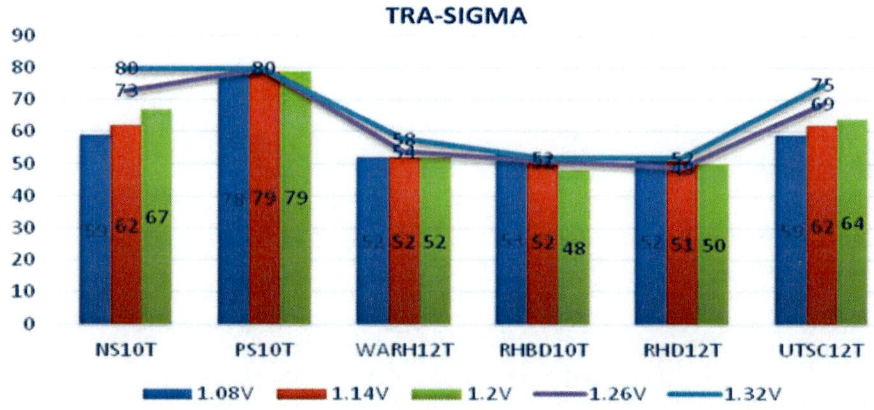

Fig. 8 TRA-SIGMA variation with voltages of the proposed WARH12T memory cell

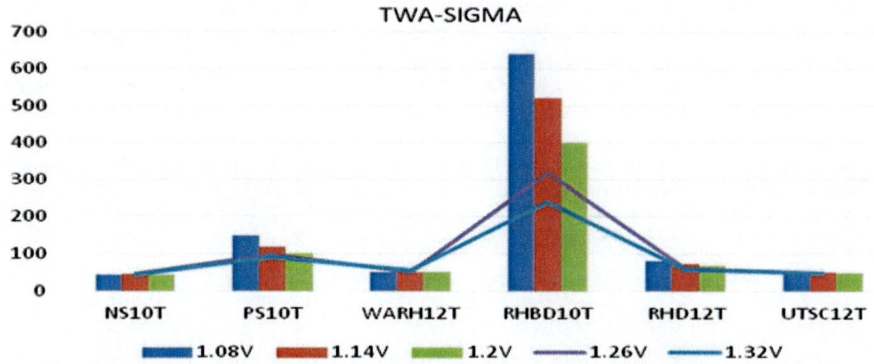

Fig. 9 TWA-SIGMA variation with voltages of the proposed WARH12T memory cell

the TWA variability of the proposed cell is similar in value to the best-performing comparison cells. Its value is slightly higher than the best-performing cells but is comparable and similar in magnitude.

4 Conclusion

This work proposed a high radiation-tolerant 12 transistor SRAM cell. The minimum charge required to change the data of the proposed cell is significantly higher than any other previously proposed cells. The radiation tolerating capacity of the proposed cell is very high compared to any other existing cell. The power requirement of the cell is also reasonably low and comparable to other cells. The read and write static noise margin of the transistor is superior to almost all preexisting structures. The hold static noise margin is also similar to comparison cells. Although the area is not good, this is a trade-off between higher radiation tolerance, write ability and area requirement. Hence, we can conclude that the proposed WARH12T cell is suitable for a very high radiation environment.

References

1. Baumann, R.C.: Radiation-induced soft errors in advanced semiconductor technologies. IEEE Trans. Device Mater. Reliab. **5**(3), 305–316 (2005)
2. Dang, L.D.T., Seo, D., Han, J.-W., Kim, J., Chang, I.-J.: A 28mn FD-SOI 4KB radiation-hardened 12T SRAM macro with 0.6 ~ 1V wide dynamic voltage scaling for space applications. In: 2018 IEEE Asian Solid-State Circuits Conference (A-SSCC), pp. 133–134. Tainan (2018)
3. Peng, C., et al.: Radiation-hardened 14T SRAM bitcell with speed and power optimized for space application. IEEE Trans. Very Large Scale Integr. (VLSI) Syst. **27**(2), 407–415 (2019)
4. Lin, D., et al.: A novel SEU tolerant memory cell for space applications. IEICE Electron. Express **15**(17), 1–8 (2018)
5. Jahinuzzaman, S.M., Rennie, D.J., Sachdev, M.: A soft error tolerant 10T SRAM bit-cell with differential read capability. IEEE Trans. Nucl. Sci. **56**(6), 3768–3773 (2009)
6. Calin, T., Nicolaidis, M., Velazco, R.: Upset hardened memory design for submicron CMOS technology. IEEE Trans. Nucl. Sci. **43**(6), 2874–2878 (1996)
7. Qi, C., Xiao, L., Wang, T., Li, J., Li, L.: A highly reliable memory cell design combined with layout-level approach to tolerant single-event upsets. IEEE Trans. Device Mater. Reliab. **16**(3), 388–395 (2016)
8. Guo, J., et al.: Design of area-efficient and highly reliable RHBD 10T memory cell for aerospace applications. IEEE Trans. Very Large Scale Integr. (VLSI) Syst. **26**(5), 991–994 (2018)

Comparative Study of InAlN and InGaN Back Barrier Layer on p-Gate/AlGaN/GaN HEMT

Shaveta, H. M. Maali Ahmed, and Rishu Chaujar

1 Introduction

In last few years due to the promising properties of gallium nitride (GaN), it is considered as the future material for devices used in space, defense, etc. applications. GaN has a large band gap and ability to operate at high temperatures as compared to other semiconductors like silicon that are usually used to fabricate devices. They also have high mobility, high saturation velocity, polarization nature and high breakdown voltage [1–3]. The common AlGaN/GaN HEMTS are normally-on due to natural induction of high-density electron gas (2DEG) in the channel. In contrast to this, normally-off devices have the benefit of positive and stable threshold voltage, high breakdown field and low on-resistance [4]. GaN-based high electron mobility transistors (HEMTs) provide exceptional performance at elevated frequency and high power electronics [4]. The reliability of the device is limited by some factors like scattering of carriers into buffer layer, which results in weakening the carrier concentration in channel and reliability [5–9]. To solve this problem, introducing a barrier (known as back barrier) region between channel and the buffer layer can reduce spilling of carriers and enhance the carrier confinement in the channel.

Several researchers reported the use of different back barrier (BB) layers. For a 65-nm gate length, Lee D. S. reported the current gain cutoff frequency (f_T) of a transistor with an AlGaN back barrier, that is 210 GHz, which is higher than that of the standard device with the same gate length [10]. T. Palacio et al. reported the use of InGaN BB layer up to f_T of 128 GHz, f_{max} of 168 GHz and improved linearity performance [11]. Y. L. Fang et al. reported the use of AlN super back barrier layer by totally replacing conventional GaN buffer and reported buffer leakage current of

Shaveta · H. M. M. Ahmed · R. Chaujar (✉)
Department of Applied Physics, Delhi Technological University, Bawana Road, Delhi, India
e-mail: chaujar.rishu@dtu.ac.in

Shaveta
Solid State Physics Laboratory, Defence Research & Development Organization, Delhi, India

A. Biswas et al. (eds.), *Microelectronics, Circuits and Systems*, Lecture Notes in Electrical Engineering 755, https://doi.org/10.1007/978-981-16-1570-2_15

10^{-5} A/mm, which is two order of magnitude lower than GaN buffer [12]. X. G. He et al. investigated the use of InAlN with 0.62–0.67 Al composition to improve carrier confinement and eliminate parasitic channel [13]. In this paper, thick GaN buffer is completely replaced by InAlN and InGaN BB layers for the AlGaN/GaN HEMT to improve the 2DEG confinement.

Here, the aim is to do comparative study between structures with InAlN BB, InGaN BB and GaN buffer for AlGaN/GaN normally-off HEMT. From different approaches, p-GaN gate is preferred due to excellent performance parameters [4]. The three structures are simulated using Silvaco TCAD software. Conduction band energy and electron concentration are investigated. The electric field distribution, output characteristics, transfer characteristics and I_{ON}/I_{OFF} ratio are studied.

The paper is planned as follows: description of structure schematic in Sect. 2, detail results and explanations in Sect. 3 followed by conclusion in Sect. 4.

2 Structure of the Proposed Device

Figure 1 highlights the structure of three different configurations of p-GaN/AlGaN/GaN HEMT that is considered for comparison. Structure I with InAlN back barrier (BB), structure II with INGaN BB and structure III with GaN buffer are taken as reference. Each structure has 150 nm GaN channel, 1 nm AlN spacer, 18 nm Al0.20Ga0.80N barrier, 70 nm-thick Mg-doped p-GaN layer having concentration of ~2×10^{19} cm^{-3}, 1 μm-thick BB (structures I and II) and 1 μm-thick GaN buffer (structure III). The gate length (*LG*) is 4 μm, gap between source and gate (*LSG*) is 4 μm and between gate and drain (*LGD*) is 15 μm. Surface state effects are reduced by a passivation layer of Si_3N_4 layer. Schottky and ohmic contact is taken at the gate and drain is ohmic terminal respectively.

Fig. 1 Schematic of enhancement-type HEMT structure

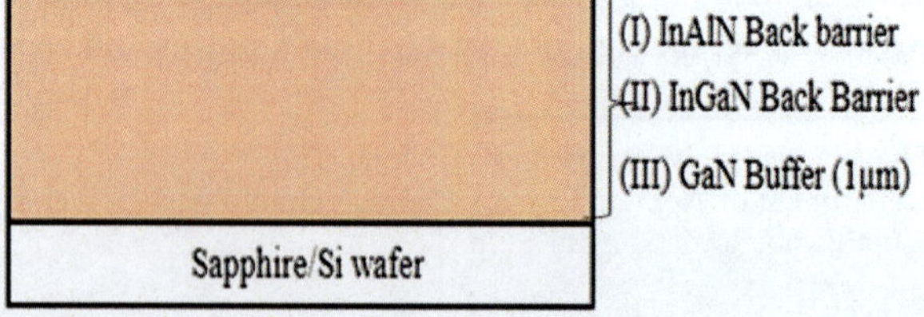

Scattering mechanism has been defined by two mobility models. The model for low field mobility is given by

$$\frac{1}{\mu(N,T)}=a\left(\frac{N}{10^{17}}\right)\left(\frac{T}{300}\right)^{-\frac{8}{2}}X\ln\left[1+3\left(\frac{T}{300}\right)^{2}\left(\frac{N}{10^{17}}\right)^{-\frac{2}{8}}+b\left(\frac{T}{300}\right)^{-\frac{8}{2}}+\frac{c}{\exp\left(\frac{1065}{T}\right)-1}\right. \tag{1}$$

where $a = 2.61 \times 10^{-4}$/V s cm^{-2}, $b = 9.8 \times 10^{-4}$/V cm^{-2} and $c = 1.7 \times 10^{-2}$/V s cm^{-2}. $\mu(N, T)$ is mobility, N is total concentration of dopant and T is temperature.

The high high-field mobility model can be specified as follows:

$$\mu(E)=\frac{\mu(N,T)+v_{sat}\frac{E(N_{1-1})}{E_c N_1}}{1+a_n\left(\frac{E}{E_c}\right)^{N_2}+\left(\frac{E}{E_c}\right)^{N_1}} \tag{2}$$

where $\mu(N, T)$ is the low field mobility, v_{sat} is saturation velocity, E is electric field. The values of E_c, a_n, N_1, N_2 are as per Ref. [14].

Polarization modeling is critical for GaN-based devices. The total charge density by polarization is defined as:

$$P_{total}=\left[P_{PE(bottom)}+P_{SP(bottom)}\right]-\left[P_{PE(top)}+P_{SP(top)}\right], \tag{3}$$

where piezoelectric and spontaneous polarization is presented by P_{PE} and P_{SP}, respectively. The difference in both polarizations will result in the formation of layer of charges at the interface.

3 Result and Discussion

As mentioned earlier, Silvaco TCAD software has been used for simulations in this paper. Important physical models like drift–diffusion transport, Fermi Dirac, low field mobility, high field mobility, Shockley-Read Hall (SRH) and polarization have been used during simulation. Following simulation parameters are taken for GaN, the band gap is 3.42 eV, the saturation velocity for electron is 2×10^7 cm/s, the spontaneous polarization is −0.034 C/m^2, the Hall electron mobility is 1900 cm^2/V s.

Figure 2 shows the conduction band profile of the three AlGaN/GaN HEMTs. It is observed that for structures I and II, the conduction band at the end of the channel is elevated as compared with the structure III. As positive polarization in both InAlN and InGaN causes conduction band edge to raise, thereby resulting in better 2DEG confinement [2].

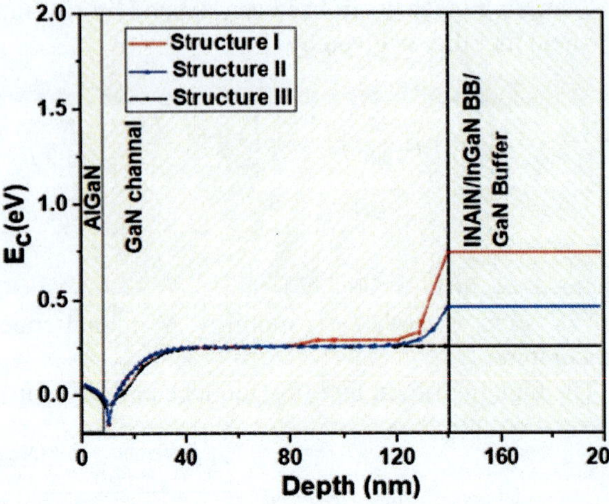

Fig. 2 Conduction band profile of HEMT structure

The effect of confinement can be seen from the simulated electron distribution profile, which is shown in Fig. 3. Structures I and II indicate electron concentration of 2.15×10^{19} cm^{-3}, 2.16×10^{19} cm^{-3}, which is more than the concentration of 8.7 $\times$ 10^{18} cm^{-3} in structure III. So the higher concentration and narrow distribution of carriers in the channel region will give better control over the device performance parameters.

Figure 4 shows the electric field distribution contours at $V_{GS} = 0$ V. An electric field is generated due to inbuilt polarization and carrier distribution at channel interface. It is seen from our simulation results, the built in electric field in region 1 of structures I and II is more concentrated at AlGaN/GaN interface, as compared with structure III. The field strength is 1.17 MV/cm, approximately one order higher

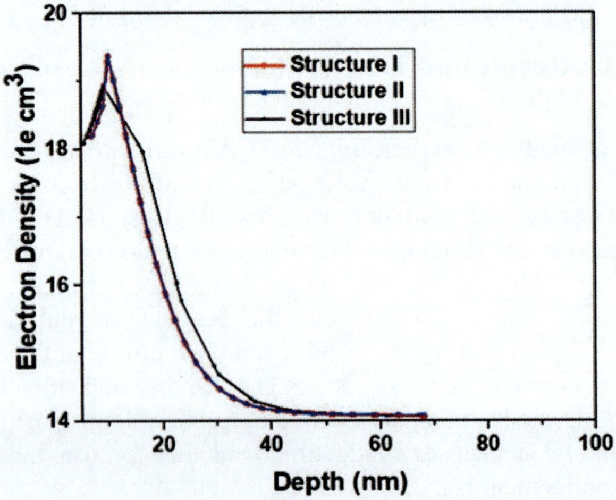

Fig. 3 Electron distribution at interface of AlGaN/GaN HEMT structure

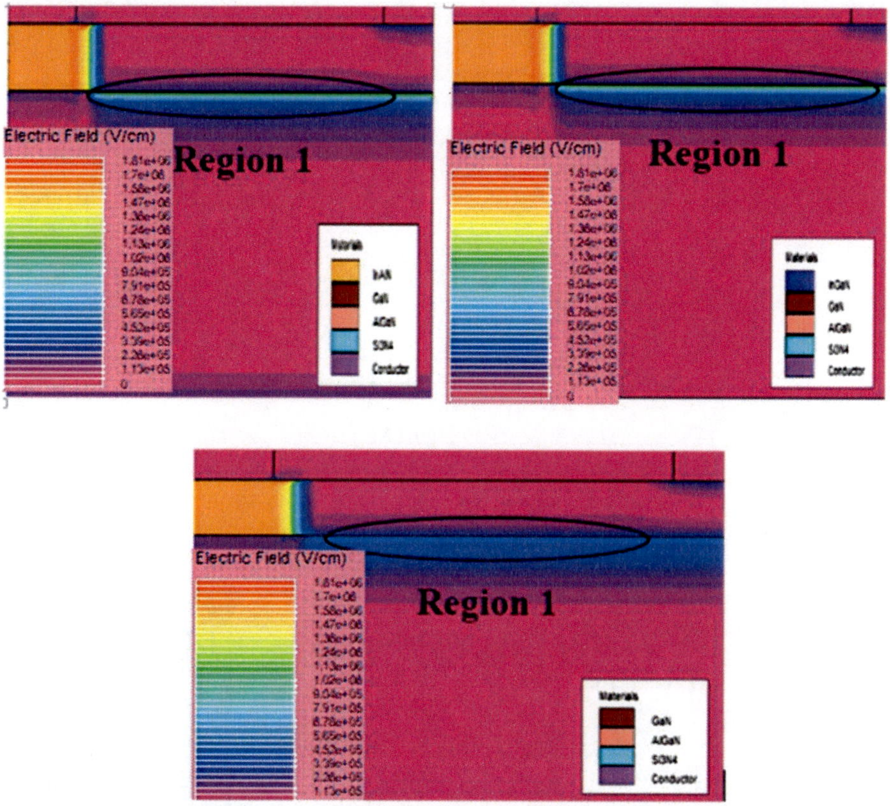

Fig. 4 Simulated electric field distribution of **a** InAlN BB **b** InGaN BB **c** GaN Buffer HEMT structures at $V_{GS} = 0$ V

than structure III. Narrower and improved carrier distribution will result in more concentrated field in the channel, which is here observed in structures I and II. Also, the formation of confined 2DEG potential well in the channel is supported by the electron distribution profile, which is shown in Fig. 2b.

Figure 5 highlights the output characteristics of different structures under investigation. The maximum drain saturation current in structures I and II is ~332 mA/mm at $V_{GS} = 6$ V, that is, 35% higher than structure III. The reason for increase in current is better carrier confinement of 2DEG in the channel. Although drain current is comparable in both back barrier layers, that might be due to almost similar electron concentration [13].

The transfer characteristics are shown in Fig. 6a. The threshold voltage is increased by ~5% with BB layer showing improved carrier concentration.

Figure 6b shows I_{ON}/I_{OFF} ratio of the three structures. I_{ON}/I_{OFF} ratio is a very important parameter as is related to switching applications. Structure I is showing a better ratio as compared with the other two structures. Although, in structure I,

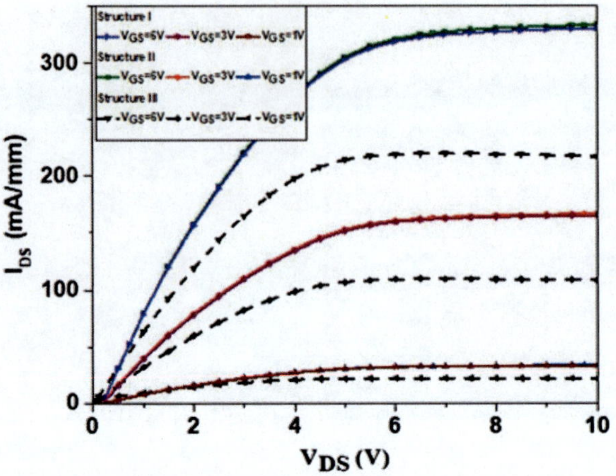

Fig. 5 I_{DS} versus V_{DS} plot of HEMTS with/without BB layer

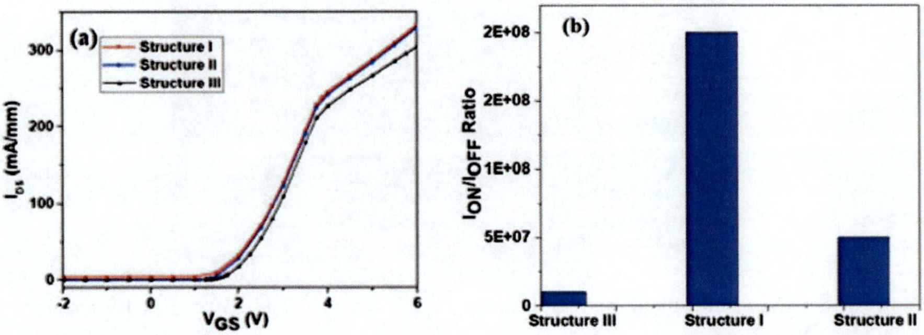

Fig. 6 **a** Transfer characteristics. **b** Switching ratio of the p-Gate/AlGaN/GaN HEMT device with/without BB layer example of a figure caption

the pinch-off current is less than structure II, it is indicating improved pinch-off characteristics as well as I_{ON}/I_{OFF} ratio. Lower band gap of InGaN BB than GaN may cause parasitic electron channel [11] that is hard to pinch-off and result in higher leakage current in subthreshold region, poorer I_{ON}/I_{OFF} ratio in the structure II.

4 Conclusion

In this paper, a comparative study of three different structures is carried out. Due to the lifting of conduction band in structures I and II (i.e. InAlN and InGaN BB, respectively), the scattering of carriers in the channel has reduced, which in turn resulted in improved 2DEG confinement. Maximum electric field in the channel is one order higher than conventional GaN buffer. The drain current is increased in structures I and II due to enhanced electron concentration in the channel. AI_{ON}/I_{OFF} ratio is higher in structures I and II as compared to structure III, which leads to better power efficiency. So, with the addition of both back barrier layers in p-GaN/AlGaN/GaN, HEMT performance is improved as compared to the conventional GaN buffer. Out of two back barrier layers, InGaN showed a poorer I_{ON}/I_{OFF} ratio due to large leakage current in the subthreshold region. Therefore, a device with InGaN back barrier may lead to higher off-state losses and reduced reliability as compared with the InAlN back barrier.

Acknowledgements This work was done with the constant support and encouragement of our guide. Heartful thanks to the Applied Physics Department, Delhi Technological University, New Delhi faculty.

References

1. Ren, F., Zolper, J.C.: Wide Band Gap Electronic Devices. World Scientific, Singapore (2003)
2. Mishar, U.K., Parikh, P., Wu, Y.: AlGaN/GaN HEMTs—an overview of device operation and applications. In: Proceedings of the IEEE 1022—1031 July 2002
3. Ambacher, O., Smart, J., Shealy, J., Weimann, N., Chu, K., Murphy, M., Chaff, W., Eastman, L., Dimitrov, R., Wittmer, L., Stutzmann, M., Rieger, W., Hilsenbeck, J.: Two-dimensional electron gases induced by spontaneous and piezoelectric polarization charges in N- and Ga-face AlGaN/GaN heterostructures. J. Appl. Phys. **85**, 3222–3233 (1999)
4. Grecoa, G., Iucolanob, F., Roccaforte, F.: Review of technology for normally-off HEMTs with p-GaN gate. Mater. Sci. Semicond. Process. **78**, 96–106 (2018)
5. Madan, J., Pandey, R., Chaujar, R.: Analysis of varied dielectrics as surface passivation on AlGaN/GaN HEMT for analog applications
6. Roccaforte, F., Fiorenza, P., Greco, G., Lo Nigro, R., Giannazzo, F., Patti, A., Saggio, M.: Challenges for energy efficient wide band gap semiconductor power devices. Phys. Status Solidi A **211**, 2063–2071 (2014)
7. Medjdoub, F., Zegaoui, M., Grimbert, B., Rolland, N., Rolland, P.-A.: Effects of AlGaN back barrier on AlN/GaN-on-silicon high-electron-mobility transistors. Appl. Phys. Express **4** (2011)
8. Kamath, A., Patil, T., Adari, R., Bhattacharya, I., Ganguly, S., Aldhaheri, R.W., Hussain, M.A., Saha, D.: Double-channel AlGaN/GaN high electron mobility transistor with backbarriers. IEEE Electron Device Lett. **33**(12) (2012)
9. Gu, Y., Chang, D., Sun, H., Zhao, J., Yang, G., Dai, Z., Ding, Y.: Theoretical study of InAlN/GaN high electron mobility transistor (HEMT) with a polarization-graded AlGaN back-barrier layer. Electronics **8**, 885 (2019)
10. Lee, D.S., Gao, X., Guo, S.: InAlN/GaNHEMTs with AlGaN back barriers. IEEE Electron Devices Lett. **32**, 617–619 (2011)
11. Palacios, T., Chakraborty, A., Keller, S., DenBaars, S.P., Mishra, U.K.: AlGaN/GaN HEMTs with anInGaN-based back-barrier. IEEE (2005)
12. Fang, Y.L., Guo, Y.M., Wang, B., Zhang, Z.R., Feng, Z.H.: High breakdown field AlGaN/GaN HEMT with AlN super back barrier
13. He, X.G., Zhao, D.G., Zhu, J.J., Chen, P., Yang, H. et al.: GaN high electron mobility transistors with AlInN back barrier. J. Alloy Comp. **662**, 16–19 (2016)

14. Ambacher, O., Majewski, J., Miskys, C., Link, A.: Hermann, pyroelectric properties of Al(In)GaN/GaN hetero-and quantum well structures. J. Phys. Condens. Matter **14**, 3399–3434 (2002)

HEMTs, MEMS and Photovoltaics

Unified Analytical Model for Charge Density and Plasmonic Waves in the Quaternary AlInGaN/AlN/GaN Heterostructures

Manju K. Chattopadhyay and Kavita Thorat Upadhyay

1 Introduction

Lately, the Terahertz (THz) sources are gaining attention due to their possible applications in spectroscopy, bio-medical imaging, and high-speed communication. Traditionally accessible terahertz suppliers included vacuum devices and cryogenic QCLs [1]. However, these devices were bulky, besides being unstable at room temperature THz signal generation.

Dyakonov and Shur theoretically envisaged that solid-state field-effect transistors could find applications as THz detectors for both resonant and non-resonant THz waves [2]. Their theory paved way for wide research on the new concept of traveling wave transistor. Various kinds of THz detectors based on solid-state devices viz. the Si metal–oxide–semiconductor field-effect transistors (MOSFETs), high electron mobility transistors (HEMTs)), and graphene FETs have been widely researched [3, 4].

As compared to the Si MOSFET, wurtzite AlGaN/GaN MODFET has superior carrier mobility and higher two-dimensional electron gas (2DEG) density, owing to the spontaneous and piezoelectric polarization effects. These effects lead to elevated responsivity and low noise equivalent power (NEP) at ambient temperature. Additionally, the manufacturing procedure is less complex as the 2DEG heterointerface is formed by spontaneous polarization without any deliberate doping. Because of these advantages, the AlGaN/GaN heterostructures are considered promising as THz detectors. Knap et al. reported an AlGaN/GaN HEMT-based THz detector with the

M. K. Chattopadhyay (✉)
School of Electronics, Devi Ahilya University, Khandwa Road, Indore 452017, MP, India

K. T. Upadhyay
Department of Electronics and Communication Engineering, IPS Academy, Institute of Engineering and Science, Indore 452012, MP, India

Department of Electronics and Telecommunication Engineering, Institute of Engineering and Technology, Devi Ahilya University, Indore 452017, MP, India

A. Biswas et al. (eds.), *Microelectronics, Circuits and Systems*, Lecture Notes in Electrical Engineering 755, https://doi.org/10.1007/978-981-16-1570-2_16

detectable frequency in the 100–600 GHz range [5]. Sun et al. demonstrated THz detection by a floating antenna-based AlGaN/GaN HEMT with elevated responsivity R_v of (3.6 kV/W) and a small NEP (40 pW/Hz$^{0.5}$) at room temperature [6]. Hou et al. also proposed a GaN HEMT with the highest R_v of 15 kV/W and the smallest value of 0.58 pW/Hz$^{0.5}$ as NEP with 0.25 μm gate length [7]. Traditionally, the AlGaN barrier layer has a constraint of the critical thickness of at least 10 nm to supply adequate 2DEG concentration. This limit blocks the frequency performance enhancement. To facilitate the increase in device frequency and to have a higher charge concentration, a thinner barrier layer is required.

The quaternary alloy semiconductor $Al_xIn_yGa_{1-x-y}N$ has clear advantages over the ternary AlGaN material. Firstly, the bandgap and lattice constants can be engineered individually, resulting in reduced defects, misfit locations, cracks, and piezoelectric field. This also allows the development of a skeletal Schottky barrier layer without significant immiscibility-associated problems in AlN, In, and GaN [8]. Secondly, with higher *Al* content, larger polarization strength can be realized because of increased spontaneous polarization in $Al_xIn_yGa_{1-x-y}N$. By using proper In and Al compositions, high 2DEG concentration in $Al_xIn_yGa_{1-x-y}N$ structures can be attained even using a thin barrier layer with better controllability by the gate voltage. Therefore, the solid-state heterostructures, consisting of a GaN channel and an AlInGaN barrier layer, are considered promising for high-power THz applications.

We propose a THz device based on AlInGaN/AlN/GaN HEMT in the present work. We present a unified analytical model for sheet charge density and plasmonic waves in quaternary AlInGaN/AlN/GaN HEMTs. This sheet charge density is used in expressions to calculate the directly proportional Plasmon frequency. Section 2 below describes our 2DEG charge density (n_s) model. Expressions for Plasmon frequency and other parameters used are given in the Sect. 2 too. Results are discussed in Sect. 3, and the present work is concluded in Sect. 4 (Figs. 1 and 2).

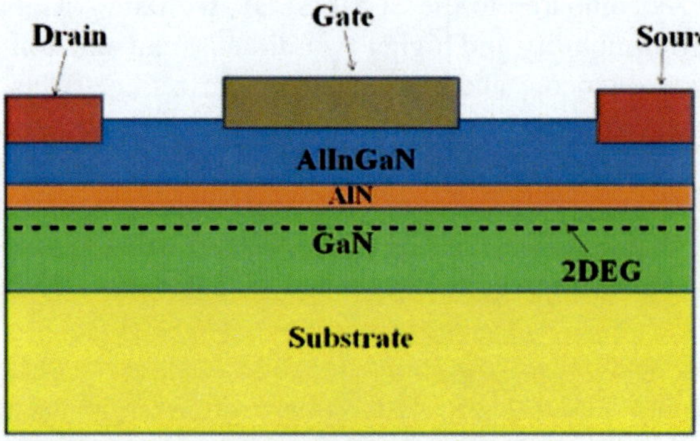

Fig. 1 Cross-sectional diagram of quaternary AlInGaN/GaN HEMT structure

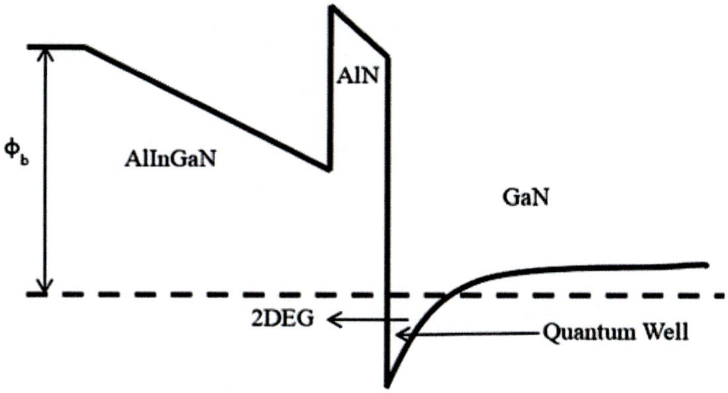

Fig. 2 Energy band diagram of the device

2 Calculations for 2DEG Charge Model

2.1 Polarization Calculations for the AlInGaN/AlN/GaN HEMT

Configuration of the 2DEG is a pertinent use of the polarization characteristics of the nitride alloy. In the literature, a good amount of valuable predictions on the polarization property of the ternary alloys is available. However, little work in literature is available with a special focus on the polarization property of quaternary AlInGaN alloy.

Structural and electronics properties can be attained by linearly interpolating the formulae based on Vegard's law. On the other hand, many structural parameters and the energy bandgap are non-linear. Therefore, in our model, we use bowing-based interpolation formulae. The bowing model requires non-linear data of applicable quantities only at the same portions of compositions [9]. In the case of the AlInGaN/GaN HEMTs, the polarization-dependent threshold voltage is given as

$$V_{th} = \varnothing_b - \Delta E_c - qN_d d_d^2/2\varepsilon - \sigma_{int}(d_d)/\varepsilon \quad (1)$$

where $\varnothing_b$ is the Schottky barrier height, ΔE_c is the conduction band discontinuity, N_d is the doping concentration, ε is the permittivity of AlInGaN, and d_d is the thickness of the AlInGaN layer,

$$\sigma_{int} = P_{GaN} - P_{AlInGaN} \quad (2)$$

and

$$P_{AlInGaN} = P_{sp}(AlInGaN) + P_{pz}(AlInGaN) \tag{3}$$

i.e., Sum of spontaneous and piezoelectric polarization.

For a practical AlInGaN/GaN heterostructure, a thin AlN layer, usually of 1 nm width, is introduced at the heterojunction to reduce the alloy disorder scattering from the AlInGaN barrier layer and to increase the charge density. We estimate its outcome by substituting the conduction band offset ΔE_c in Eq. (1) by our modified conduction band offset as

$$\Delta E_{c\_mod} = \Delta E_{c\_AlN/GaN} - \Delta E_{c\_AlInGaN/AlN} + \Delta E_{c\_AlN} \tag{4}$$

Here, we have taken

$$\Delta E_{c\_AlN} = \frac{q^2 d_{AlN}}{\varepsilon_{AlN}}\left(\frac{\sigma_{AlN}}{q} - n_s\right) \tag{5}$$

Various other terms are as defined in Table 1.

2.2 *Carrier Density Model*

The sheet charge density n_s can be obtained by solving the 1D Poisson's equation and can be represented as

$$n_s = \varepsilon\left(V_g - V_{th} - V_k - E_F/q\right)/q\mathrm{d} \tag{6}$$

where d is the sum of d_d and the length of the AlN layer, V_k is the channel voltage at distance k, and E_F is the Fermi level.

An analytical expression for n_s as a function of V_g can be expressed as [10]

$$n_s = \left[AV_{go}/(1+B)\right]\left[\left(1 - A^{2/3}\gamma_0\right)/(1+B)^{2/3}V_{go}^{1/3}\right] \tag{7}$$

$A = \varepsilon/qd$ and $B = A/qD$, and D is the two-dimensional density of states inside the quantum well at the heterointerface.

The unified expression for n_s including the contribution of charge carriers in the barrier layer is given as [11]

$$n_{s,unified} = \frac{2k_B T A\ln\left(1 + exp^{\left(\frac{qV_{go}}{2k_BT}\right)}\right)}{q\left(\frac{AV_{go}}{n_{s,AboveVth}} + \frac{A}{D}exp^{\left(\frac{-qV_{go}}{2k_BT}\right)}\right)} \tag{8}$$

here $V_{go} = V_g - V_{th}$.

Table 1 Material specific parameters for the calculations

Parameters	Values/Details
Polarization charge across the AlN/GaN interface	σ_{AlN}
Thickness of the AlN layer	d_{AlN}
Conduction band offset at the AlN/GaN interface	$\Delta E_{c\_AlN/GaN}$
Conduction band offset at the AlInGaN/GaN interface	$\Delta E_{c\_AlInGaN/AlN}$
Permittivity of the AlN layer	ε_{AlN}
Spontaneous polarization	$P_{sp}(AlInGaN) = x \cdot P_{sp}(AlN) + y \cdot P_{sp}(InN) + z \cdot P_{sp}(GaN) + b_{AlGaN} \cdot z \cdot + b_{InGaN} \cdot y \cdot z + b_{AlInN} \cdot x \cdot y + b_{AlInGaN} \cdot x \cdot y \cdot z$
$b_{AlGaN}, b_{InGaN}, b_{AlInN}$ are bowing parameters of ternary alloy	$b_{AlInGaN} = 27P_{sp}(Al_{0.33}In_{0.33}Ga_{0.33}N) - 9(b_{AlGaN} + b_{InGaN} + b_{AlInN}) - 3(P_{sp}(AlN) + P_{sp}(InN) + P_{sp}(GaN))$
Basal strain defined after Vegard's interpolation of lattice parameter	$\eta(U_x V_y Ga_z N) = \frac{[x(a^{GaN} - a^{UN}) + y(a^{GaN} - a^{VN})]}{[xa^{UN} + ya^{VN} + za^{GaN}]}$
Piezoelectric polarization parameters	$P_{pz}(AlInGaN) = x \cdot P_{pz}(AlN, \eta) + y \cdot P_{pz}(InN, \eta) + z \cdot P_{pz}(GaN, \eta)$ $P_{pz}(AlN, \eta) = -1.808\eta + 5.624\eta^2 for \eta < 0$ $P_{pz}(AlN, \eta) = -1.808\eta - 7.888\eta^2 for \eta > 0$ $P_{pz}(GaN, \eta) = -0.918\eta + 9.541\eta^2$ $P_{pz}(InN, \eta) = -1.373\eta + 7.559\eta^2$
Energy bandgap	$E_g(GaN) = 3.4\text{eV}, E_g(AlN) = 6.2\text{eV}, E_g(InN) = 0.7eV$

2.3 *Model for Drain Current*

The model for drain current may be formulated by calculating the current along the channel.

$$I_d = q\mu_0 W n_{s,unified} v_s \tag{9}$$

where μ_0 is the low vertical field mobility, *W* is the width of the device, and v_s is the drift velocity.

Integrating from source to drain gives a simple analytical model of the drain current as

$$I_d = \left(\frac{q\mu_{eff}W}{L_g}\right)\left\{\left[\frac{(A+D)\left(n_s^2 - n_D^2\right)}{2AD}\right] + \left(\frac{2}{5}\gamma_0\left(n_s^{5/3} - n_D^{5/3}\right)\right) + \frac{k_B T(n_s - n_D)}{q}\right\} \quad (10)$$

where n_s and n_D are the electron charge densities at the source and drain ends, respectively, L_g is the gate length, and the rest of the symbols are defined as done conventionally. The calculations of n_D and saturation voltage V_{sat} are given in our previous work [10].

The second-order effects such as drain-induced barrier lowering (DIBL), channel length modulation, and self-heating effects (SHE) are neglected here but can be included in I_d–V_d characteristics by modifying the Eq. (4), in a procedure similar to our earlier work [12].

2.4 Dispersion Relations for Plasma Waves in 2D Gated Devices

In two- or one-dimensional plasma, the frequency of plasma oscillations, ω_p depends on the plasma wave vector k. Theoretical modeling of a 2D plasma leads to the following dispersion relationship [13]:

$$\omega_p = \sqrt{\frac{q^2 n_s k}{2m^* \varepsilon_0 \varepsilon_{GaN}}},\ k = \sqrt{\frac{qV_{go}}{m^*}} k_n \text{ for } kd \ll 1$$

$$(k_n)_{gated} = \frac{(2n-1)\pi}{L_g}, \text{ where } n = 1, 2, 3\ldots$$

The plasma waves propagate with much greater velocities than the electron drift velocities. Hence, the plasma wave frequency for the fundamental harmonic for short-channel solid-state devices is in the THz range. Other relations used for various parameters related to plasmon oscillations are listed in Table 2.

Table 2 Relationship used for various parameters related to plasmon oscillations

Parameter with units	Relationship
Characteristic length for the ac voltage decay from the source [15]	$L_o = s\sqrt{\tau/m}$
Electron effective mass in AlInGaN layer	$m^* = (0.35x + 0.1y + 0.2z)m_0$
Fermi velocity [16]	$v_F = \frac{h}{2\pi}\frac{1}{m^*}\sqrt{2\pi n_s}$
Electron plasma velocity [16]	$S = \sqrt{\frac{q^2 n_s d}{\epsilon m^*}}$
Fundamental plasma frequency [16]	$f = \frac{S}{4L_g}$

Plasma wave electronic devices have a major benefit that they can be tuned by applying bias voltage and are able to modulate at very high frequencies viz. 200 GHz and more [14]. This makes quaternary GaN heterostructures interesting devices for probable employability in high-power, ultrahigh-speed wireless communications used in the ultrahigh-resolution broadcast.

3 Results and Discussion

Figure 3 shows the comparison of our calculated output characteristics and experimental data published by Hwang et al. [8]. Values of various parameters used are given in Table 3. High drain currents of 1589 A/mm, 1278 A/mm, 865 A/mm, and

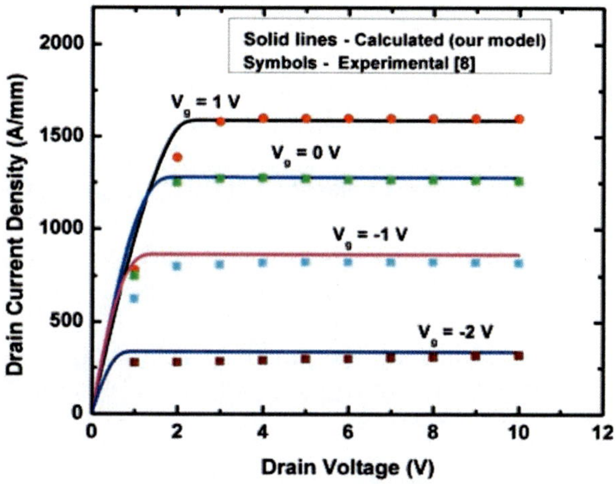

Fig. 3 I_d–V_d characteristics of heterostructure at different gate voltages V_g

Table 3 Model and device parameters used in calculations for comparison in figure

Parameter with units		Values employed in our computations
L_g (μm)		400 nm
W (μm)		1 μm
μ_0 (cm^2/V-s)		1621
Fitting parameters	a_1 (1/V)	-40×10^{-3}
	a_2 (1/V^2)	0.32
	a_3(1/V)	5×10^{-5}
Transition width parameter	α, (unit less)	1.2
R_s (Ω)		0.6
R_d (Ω)		0.6

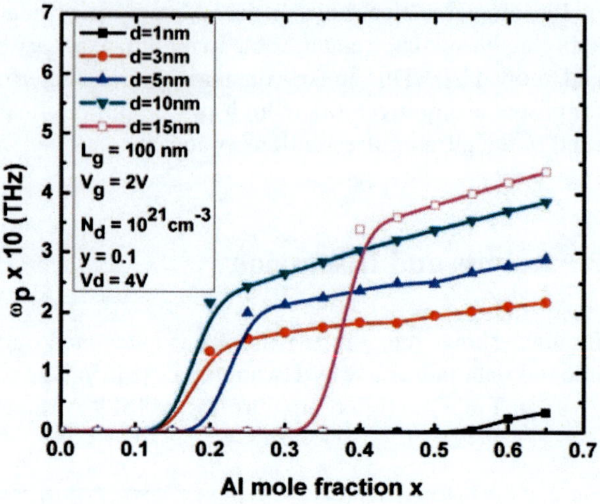

Fig. 4 Plasma frequency ω_p versus *Al* mole fraction *x*

337 A/mm are obtained at gate voltages 1 V, 0 V, −1 V, and −2 V, respectively. By choosing suitable fitting parameters, we obtained a good match with the experimental result.

The relationship of plasma frequency ω_p with *x,* the *Al* mole fraction, is depicted in Fig. 4 for $L_g = 100$ nm. A higher frequency range is observed for higher values of *d*. The frequencies observed are higher by an order as compared to calculations done for AlGaN/GaN HEMTs by Rabbaa et al. [17]. ω_p achieved values more than 40 THz for $d = 15$ nm and $L_g = 100$ nm.

Figure 5 illustrates the variation of plasma frequency ω_p with doping density at different gate voltages applied. A linear relationship of ω_p is seen with the doping

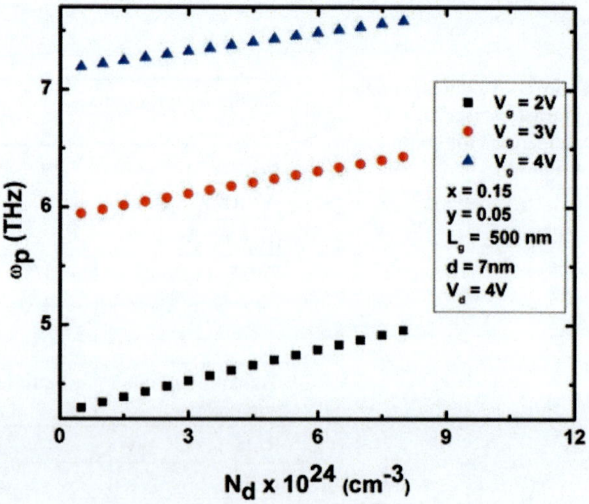

Fig. 5 Plasma frequency versus doping density for different gate voltages V_g

density N_d and the V_g. The relationship of Plasma and Fermi velocities with 2DEG is shown in Fig. 6. We infer that the plasma wave velocity reaches a much higher value of 27.6×10^5 m/s as compared to the Fermi velocity of 9.4×10^5 m/s for a sheet charge concentration of 10^{17} cm^{-3}.

Change in calculated plasma frequency ω_p versus V_g for different *In* mole fractions *y* is illustrated in Fig. 7 with *Al* mole fraction *x* fixed at 0.3. It is observed that the lower *y* values give higher plasma frequencies. Figure 8 depicts the variation of fundamental plasma frequency f with the gate length L_g at a drain voltage of 4 V for gate voltages from 2 to 5 V. We see that f rises exponentially with a decrease in L_g. Although we attain great THz plasmon frequencies by lessening the L_g, there are certain limitations

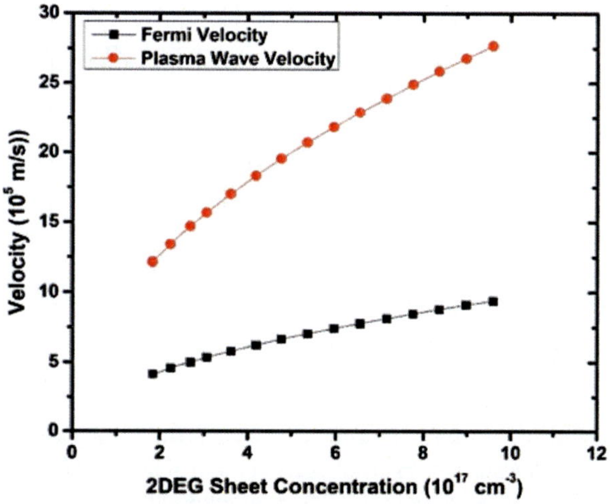

Fig. 6 Plasma velocity S and Fermi velocity v_F versus 2DEG sheet concentration

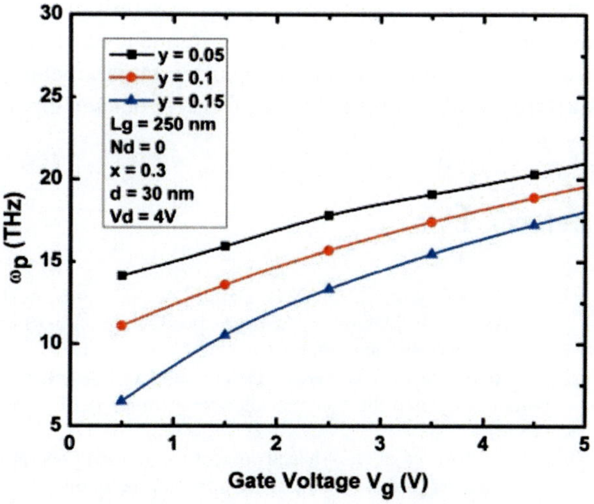

Fig. 7 Calculated plasma frequency ω_p versus gate voltage for different Indium mole fractions y

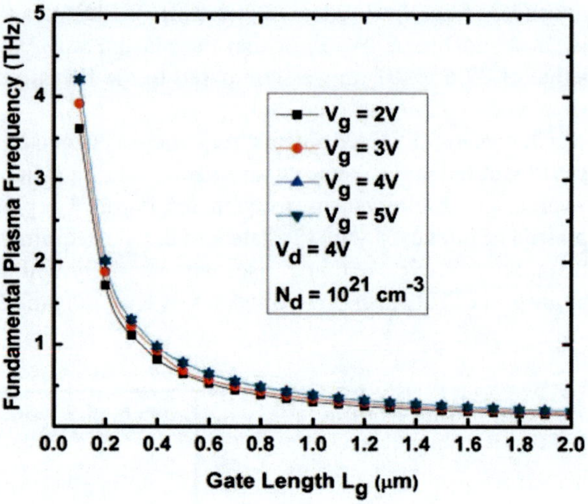

Fig. 8 Fundamental plasma frequency f versus gate length L_g for various gate voltages

with very short gate lengths while designing the HEMTs. Therefore, one has to ascertain the minimum critical gate length for optimized operation.

4 Conclusion

A unified analytical model for calculating sheet charge density and plasma frequency of Terahertz waves in quaternary AlInGaN/AlN/GaN HEMTs is presented. Calculated results using our model matched well for I_d–V_d characteristics of experimental data given in Ref. [8]. We observed that due to higher electron concentration accumulation in quaternary alloys as compared to ternary alloys such as AlGaN, the plasmon oscillations introduced have higher frequency by almost an order. Our unified analytical model gives an overview of different parameter relationships for exploring the quaternary alloy-based HEMTs for THz device applications.

References

1. Dunn, A., Poyser, C., Dean, P., Demić, A., Valavanis, A., Indjin, D., Salih, M., Kundu, I., Li, L., Akimov, A., Davies, A.G.: High-speed modulation of a terahertz quantum cascade laser by coherent acoustic phonon pulses. Nat. Commun. **11**(1), 1–8 (2020)
2. Dyakonov, M.I., Shur, M.S.: Detection, mixing, and frequency multiplication of terahertz radiation by two-dimensional electronic fluid. IEEE Trans. Electron Dev. **43**(10), 380–387 (1996)
3. Ahi, K.: Review of GaN-based devices for terahertz operation. Opt. Eng. **56**(9), 090901 (2017)
4. Božanić, M., Sinha, S.: Emerging transistor technologies capable of terahertz amplification: a way to re-engineer terahertz radar sensors. Sensors **19**(11), 2454 (2019)

5. Knap, W., Dyakonov, M., Coquillat, D., Teppe, F., Dyakonova, N., Łusakowski, J., Karpierz, K., Sakowicz, M., Valusis, G., Seliuta, D., Kasalynas, I.: Field effect transistors for terahertz detection: physics and first imaging applications. J. Infrared Millim. Terahertz Waves **30**(12), 1319–1337 (2009)
6. Sun, J.D., Sun, Y.F., Wu, D.M., Cai, Y., Qin, H., Zhang, B.S.: High-responsivity, low-noise, room-temperature, self-mixing terahertz detector realized using floating antennas on a GaN-based field-effect transistor. Appl. Phys. Lett. **100**(1), 013506 (2012)
7. Hou, H., Liu, Z., Teng, J., Palacios, T., Chua, S.J.: A sub-terahertz broadband detector based on a GaN high-electron-mobility transistor with nanoantennas. Appl. Phys. Express **10**(1), 014101 (2016)
8. Hwang, J.H., Kim, S.M., Woo, J.M., Hong, S.M., Jang, J.H.: GaN HEMTs with quaternary In0. 05Al0. 75Ga0. 2N Schottky barrier layer. Physica Status Solidi A **213**(4), 889–892 (2016)
9. Fiorentini, V., Bernardini, F., Ambacher, O.: Evidence for nonlinear macroscopic polarization in III–V nitride alloy heterostructures. Appl. Phys. Lett. **80**(7), 1204–1206 (2002)
10. Upadhyay, K.T., Chattopadhyay, M.K.: Al composition and $Al_xIn_yGa_zN$ layer thickness dependent new analytical model for I-V characteristics of $Al_xIn_yGa_zN$/GaN HEMTs. **19**(2), 205–208 (2019)
11. Swamy, N.S., Dutta, A.K.: Analytical models for the 2DEG density, AlGaN layer carrier density, and drain current for AlGaN/GaN HEMTs. IEEE Trans. Electron. Dev. **65**(3), 936–944 (2018)
12. Chattopadhyay, M.K., Tokekar, S.: Thermal model for dc characteristics of ALGaN/GaN hemts including self-heating effect and non-linear polarization. Microelectron. J. **39**(10), 1181–1188 (2008)
13. Stern, F.: Polarizability of a two-dimensional electron gas. Phys. Rev. Lett. **18**(14), 546 (1967)
14. Otsuji, T., Shur, M.: Terahertz plasmonics: good results and great expectations. IEEE Microwave Mag. **15**(7), 43–50 (2017)
15. Liu, Shur, M.: An efficient TCAD model for TeraFET detectors. In: IEEE Radio and Wireless Symposium (RWS) (2019)
16. Shur, M.: AlGaN/GaN plasmonic terahertz electronic devices. J. Phys. Conf. Ser. **486**, 012025 (2014)
17. Rabbaa, S., Stiens, J.: Charge density and plasmon modes in a triangular quantum well model for doped and undoped gated AlGaN/GaN HEMTs. J. Phys. D Appl. Phys. **44**, 32 (2011)

Studies on Performances of Copper Oxide Nanoparticles from *Catharanthus Roseus* Leaf Extract

K. A. Khan, M. Shaiful Islam, Md. Abdul Awal, M. N. Islam Khan, and A. K. M. Atique Ullah

1 Introduction

The CuO Nanoparticles from *Catharanthus Roseus* leaf extract is an innovative idea [1–4]. Although it has been conducted MnO2 NPs from *Bryophillim pinnatum* leaf extract by our research group previously[5–10] It has been conducted the different characterizations like XRD, GC–MS, SEM, FESEM, FTIR, EDX, DSC, VSM, TGA, etc. The CuO nanoparticles are stable, robust, and have a longer shelf life compared to organic, antimicrobial agents [11–13]. There are a few strategies for the union of CuO NPs, which incorporate sol–gel process, co-precipitation, sonochemical method, hydrothermal techniques, non-aqueous synthesis, ultrasound irradiation, micro-emulsion method, electrochemical and microwave irradiation synthetic route, and so on[14–16]. The main aim of our research work on CuO Nanoparticles (NPs) is to use for electricity generation system by using electrochemical cell. The *Catharanthus Roseus* leaf is also a medicinal leaf [17–21]. It has a great medicinal value. There are some countries where it is used as a medicinal value. The biogenic amalgamation includes the blend of NPs utilizing plants and microorganisms or their concentrates [22, 23]. The microbial combination is a tedious and to some degree dangerous strategy, since it includes the upkeep of microbial societies under advanced, septic research center conditions. Developing a facile and green method for synthesizing

K. A. Khan (✉) · M. Shaiful Islam
Department of Physics, Jagannath University, Dhaka 1100, Bangladesh

Md. A. Awal
Department of Chemistry, National University, Gazipur, Bangladesh

M. N. I. Khan
Nanoscience and Technology Research Laboratory, Atomic Energy Centre, Atomic Energy Commission, Dhaka 1000, Bangladesh

A. K. M. A. Ullah
Nanoscience and Technology Research Laboratory, Chemistry Division, Atomic Energy Centre, Bangladesh Atomic Energy Commission, Dhaka 1000, Bangladesh

A. Biswas et al. (eds.), *Microelectronics, Circuits and Systems*, Lecture Notes in Electrical Engineering 755, https://doi.org/10.1007/978-981-16-1570-2_17

CuO Nanoparticles (NPs) is of importance and still a challenge for materials scientists [24, 25]. The *Catharanthus roseus* leaves possess biomolecules such as terpenoid, carotenoid, sterol, alcohol, and organic acid, which could be used as reducing agent to react with Cu^{2+} ions as scaffolds to direct the formation of the CuO Nanoparticles (NPs) in a solution that have been extensively explored for possible medical applications [26]. The CuO Nanoparticles (NPs) from *Catharanthus roseus* leaves provide a particularly useful platform and demonstrate unique properties with potentially wide-ranging therapeutic applications [27].

2 Methodology

CuO NPs were synthesized via a facile green synthesis route where ($Cu(CH_3COO)_2$-H_2O) were used as a precursor and *Catharanthus roseus* leaf extract was used as a source of reducing and capping agents. To synthesis CuO NPs, the *Catharanthus roseus* leaf extract was added to an aqueous mixture of Cu^{2+} acetate at a 1:1 M ratio. The chemical reactions for CuO NPs are given by the following:

Catharanthus roseus + H_2O + $Cu^{2+}_{(aq)}$ $\xrightarrow{\text{Stirring}}$ [*Cathararanthus roseus*/Cu^{2+}]
$\rightarrow$ [*Cathararanthus roseus*/Cu^{2+}] [*Cathararanthus roseus*/CuO] $\downarrow$(s) + H_2O (aq)

The method of *Catharanthus Roseus* leaf extract preparation for getting CuO Nanoparticles (NPs) is shown in Fig. 1. From Fig. 1, it is found that the finished product was *Catharanthus Roseus* leaf extract (*C.R.*L. extract).

The method of CuO Nanoparticles (NPs) preparation using *Catharanthus Roseus* leaf extract is shown in Fig. 2. Finally, it is found (from Fig. 2) that the finished product was CuO Nanoparticles (NPs) using *Catharanthus Roseus* leaf extract (*C.R.*L. extract).

Fig. 1 Method of leaf extracts preparation

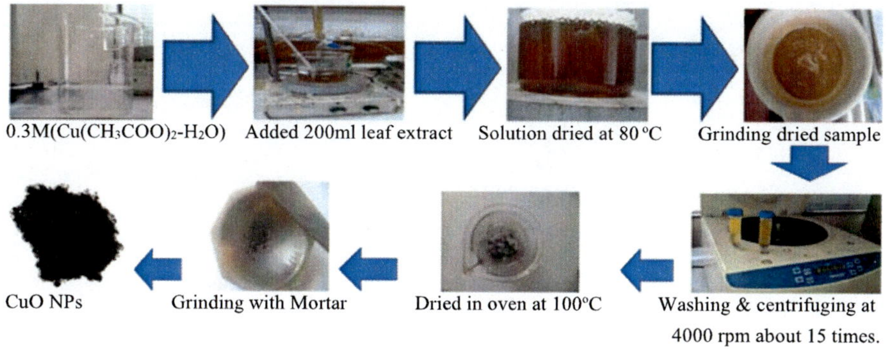

Fig. 2 Methods for sample preparation of CuO NPs

3 Results and Discussion

Table 1 shows the amount inpercentage (%) of functional group of *Catharanthus roseus* leaf extracts. It is found that the maximum functional group present in *Catharanthus roseus* leaf extract was Alcohol (42.64%) and the minimumfFunctional group present in *Catharanthus roseus* leaf extract was Carotenoid (1.82%).

3.1 Structural Analysis Using XRD

It is shown (in Fig. 3) that the diffraction peaks of synthesized CuO NPs are assigned to the crystal planes of (110), (−111), (111); (−202), (020), (202); (−113), (−311), and (220), respectively. The analyzed diffraction peaks were matched well with the standard magnetite XRD patterns with JCPDS file no: 05-0661, which declared the crystallographic system of the monoclinic structure.

Table 1 Functional group present in *Catharanthus roseus* leaf extracts

Functional group	Amount (%)
Alcohol	42.64
sterol	23.03
Alkane	3.43
Ketone	10.14
Terpenoid	2.67
Carboxylic acid	2.43
Carotenoid	1.82

Fig. 3 XRD patterns of CuO Nanoparticles

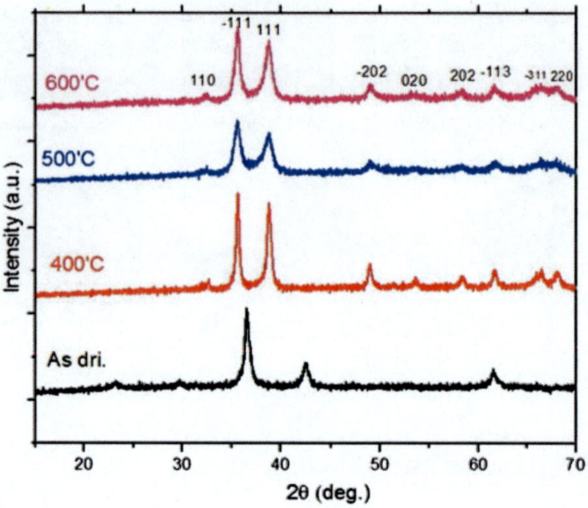

3.2 Structural Parameter of CuO NPs

The structural parameter for different calcination temperatures at 400 °C, 500 °C, and 600 °C is shown in Table 2. The maximum crystallite size, D(nm) was 30.77 nm at 400 °C and the minimum crystallite size, D(nm) was 13.25 nm at 500 °C. The maximum dislocation density, δ(line/nm^2 × 10^{-3}) was 5.70 at 500 °C and the minimum dislocation density, δ(line/nm^2 × 10^{-3}) was 1.06 at 400 °C. The maximum microstrain, Ɛ(×10^{-3}) was 1.34 at 500 °C and the minimum microstrain, Ɛ(×10^{-3}) was 0.85 at 400 °C.

It is shown (in Fig. 4) that the average crystallite size decreases from 400 °C to 500 °C with increasing calcination temperature and then increases with temperature. It is shown (in Fig. 5) that the dislocation density increases with increasing calcination temperature up to 500 °C and then decreases with temperature.

Table 2 Structural parameter for different calcination temperatures

Temperature(°C)	Crystallite Size, D(nm)	Dislocation density, δ(line/nm^2 × 10^{-3})	Microstrain, Ɛ(×10^{-3})
400	30.77	1.06	0.85
500	13.25	5.70	1.34
600	21.78	2.11	1.18

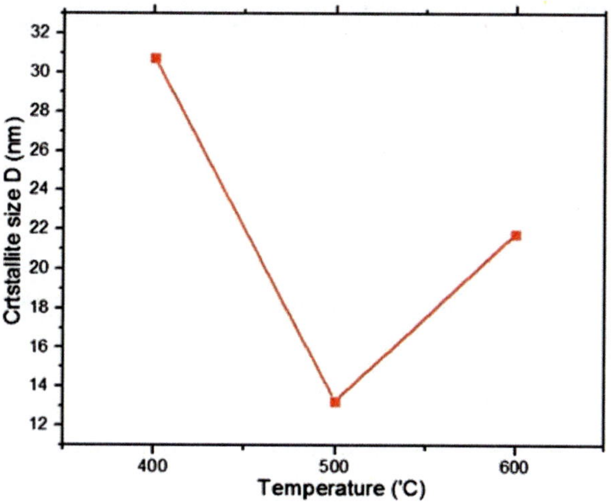

Fig. 4 Calcination temperature versus crystallites size

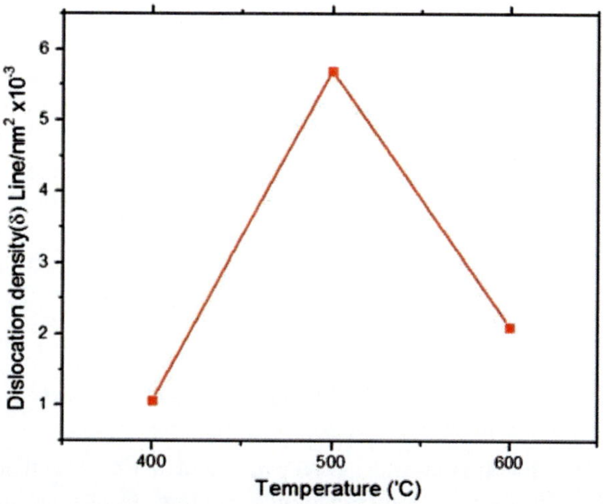

Fig. 5 Calcination temperature versus Dislocation density (δ)

3.3 *Williamson–Hall Method*

Williamson–Hall X-ray line broadening analysis provides a method of finding an average size of coherently diffracting domains and strain. Strain-induced peak broadening arises due to crystal imperfection; and distortion, which had been calculated using the relation :

$$\varepsilon = \frac{\beta}{4\tan\theta} \tag{1}$$

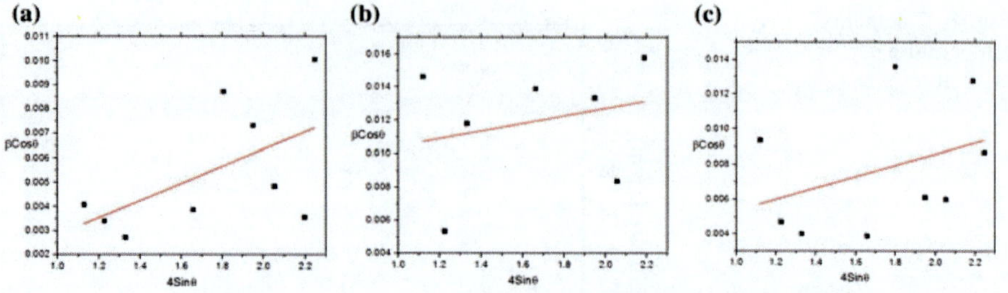

Fig. 6 Slope of the plots provides the value of strain. The slope for 400'C, 500'C, and 600'C are 3.53×10^{-3}, 2.24×10^{-3}, and 3.23×10^{-3}, respectively

To estimate microstrain from the X-ray diffraction pattern, Williamson and Hall [9] proposed a modified Scherrer's formula as follows:

$$\beta \cos\theta = \frac{\mathrm{k}\lambda}{\mathrm{D}} + 4\varepsilon \sin\theta \quad (2)$$

The equation represents the linear plot of βhklcosθ_ against 4 sinθ for the samples calcined at (a) 400 °C, (b) 500 °C, and (c) 600 °C temperatures displayed in Fig. 6. The slope of the plots provides the values of strain (ε) [9].

It is shown (in Fig. 6) that the plots provide the value of the strain, which varies with temperatures. The value of the slope of the strain was 3.53×10^{-3}, 2.24×10^{-3}, and 3.23 × 10–3 at 400'C, 500'C, and 600'C, respectively. It is found that the maximum value of the slope of the strain was 3.53×10^{-3} at 400 °C and the minimum value of the slope of the strain was 2.24×10^{-3} at 500 °C.

3.4 *Surface Morphology Analysis: SEM and FESEM*

The SEM images of CuO NPs calcined at 400 °C and 500 °C are shown in Fig. 7a and Fig. 7b. The SEM image shows that the samples are in the nano range. The FESEM

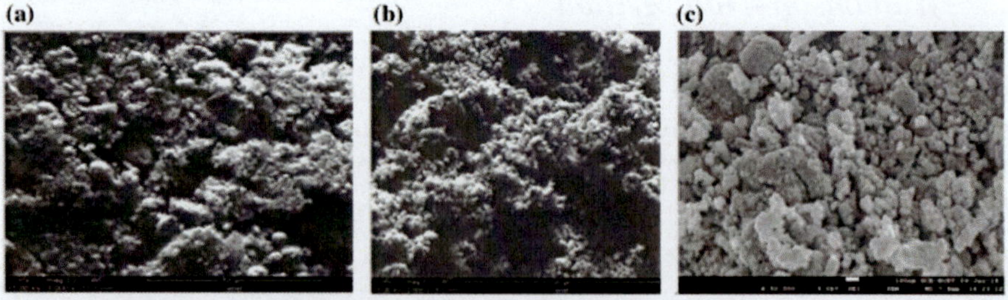

Fig. 7 Surface morphology analysis using SEM and FESEM

image of CuO Nanoparticles (NPs) calcined at 600 °C is shown in Fig. 7c. The image shows that the surface morphology of the Nanoparticles (NPs) is homogeneous and almost spherical in shape.

3.5 *Functional Group Analysis of Nanoparticles Using FTIR Technique*

Table 3 shows the functional group analysis of Nanoparticles (NPs) using the FTIR technique. It is also found from Table 3 that the maximum peak position(cm^{-1}) of the functional group was 3460 for (-OH) and the minimum peak position(cm^{-1}) of the functional group was 515 for (Cu–O).

It is shown (in Fig. 8) that the stretching vibration of –OH functional group is at thepPeak position of 3460 cm^{-1}. Whereas, the antisymmetric stretching vibration of $–CH_2–$ is in naphthalic or aliphatic at the peak position of 2925 cm^{-1}. Again, it is also shown that the stretching vibration of –OH functional group is in polyols at the

Table 3 Table for FTIR analysis of CuO NPs

Functional Group	Peak Position(cm^{-1})	Attribution
– OH	3460	Stretching vibration of –OH functional Group
– CH_2–	2925	Antisymmetric stretching vibration of $–CH_2–$ in naphthalic or aliphatic
– OH	1416	Stretching vibration of –OH in polyols
C–N	1050	Stretching vibration of C-N in amines
Cu–O	515	Stretching vibration of Cu–O bond

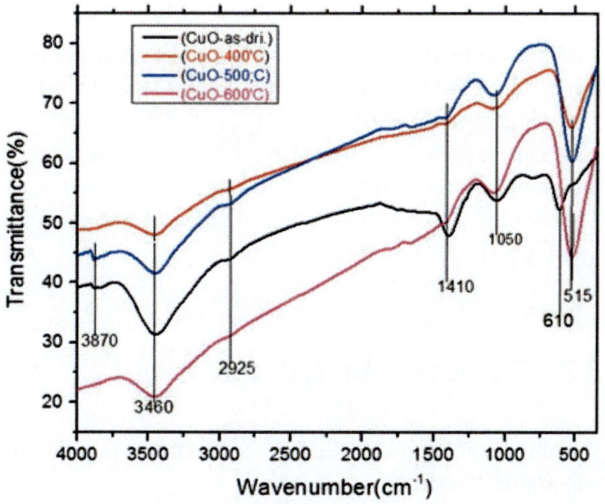

Fig. 8 FTIR analysis of CuO NPs

Table 4 Elemental analysis

Elements	Mass (%)
Copper (Cu)	40.36
Oxygen (O)	20.52
Carbon (C)	39.12

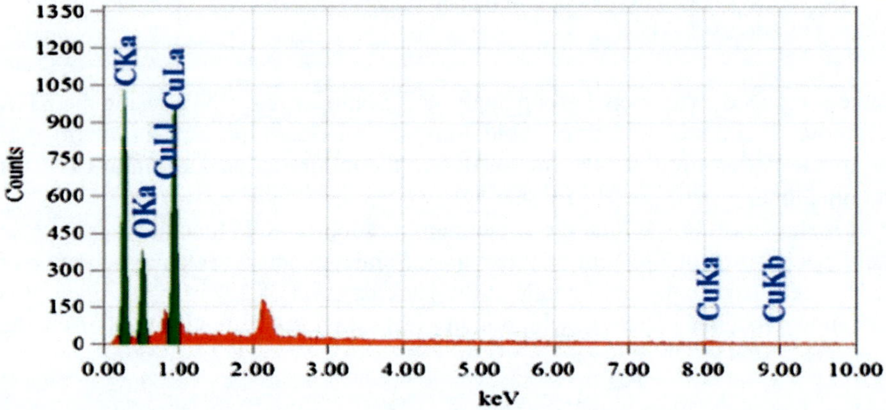

Fig. 9 Elemental analysis of CuO NPs calcined at 600 °C using EDX

peak position of 1416 cm^{-1}, whereas, the stretching vibration of C-N is in amines at the peak position of 1050 cm^{-1}. Finally, it is shown that the stretching vibration of Cu–O bond is at the peak position of 515 cm^{-1}.

3.6 Elemental Analysis of CuO NPs

Table 4 shows the Elemental Analysis of CuO Nanoparticles (NPs). It is shown that the mass (%) of Copper (Cu), Oxygen (O), and Carbon (C) are 40.36%, 20.52%, and 39.12%, respectively.

It can be seen that (from Fig. 9) no coercivity or remanence could be observed for the sample, suggesting the diamagnetic properties of the magnetite nanoparticles. This can be ascribed to the small size of nanoparticles which were smaller than the diamagnetic critical size.

3.7 Elemental Analysis and Magnetization of CuO NPs

It is shown (in Fig. 10) that the saturation magnetization of the nanoparticles (NPs) was 0.00125 emu/g.

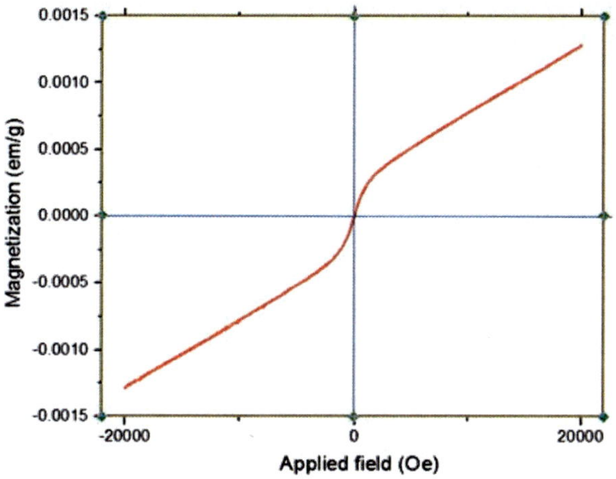

Fig. 10 Magnetization measurement: VSM image of CuO NPs

3.8 Thermal Analysis of Nanoparticles Using TGA

In the TGA plot (from Fig. 11), the total weight losses are observed 13.33% up to 1100 °C. The weight losses are observed 2.36% up to 700 °C and 10.81% up to 800–1000 °C flat line stabilizes the pure CuO NPs with no impurities.

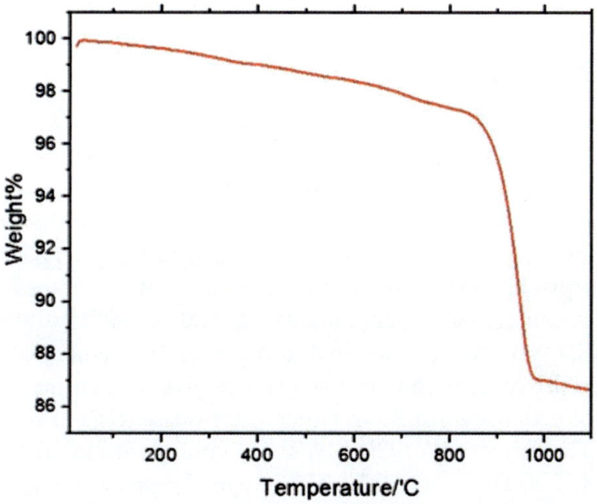

Fig. 11 TGA plot of CuO NPs calcined at 600 °C

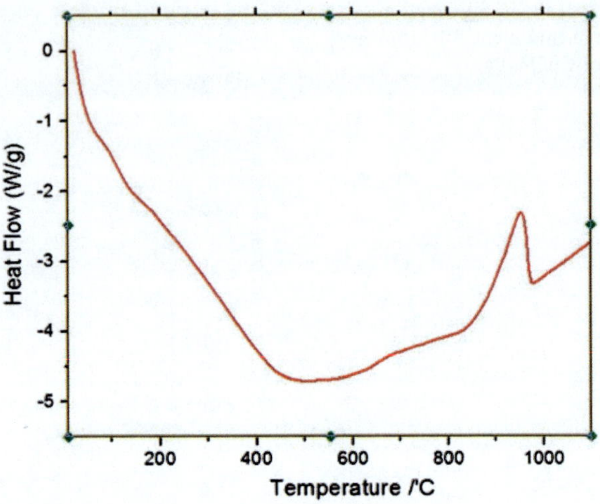

Fig. 12 DSC plot of CuO NPs calcined at 600 °C

3.9 Thermal Analysis of Nanoparticles Using DSC

The DSC curve shows (from Fig. 12) a broad exothermic dip in between room temperature and 500 °C. There is also a sharp dip observed at around 950 °C, which might be due to crystallization and condensation to the anatase phase.

4 Conclusions

The XRD pattern of the average crystallite size was estimated for the synthesized spherical monocyclic crystalline typed CuO NPs as 21.96 nm. NPs synthesized with various calcination temperature also confirmed their crystallinity by XRD analysis and it was found that all of them have pure crystal phase with a spherical monocyclic crystalline structure, where the crystallites size of the NPs decreased with the increase of calcination temperature 400 to 500 °C and crystallite size increases with 600 °C. EDX spectra confirmed the presence of elemental components of the CuO NPs. FTIR spectroscopy showed the involvement of biomolecules present in the extract of *Cathranthus roseus*, which were verified in the synthesizing process of CuO NPs. The formation of CuO NPs was confirmed due to the noticeable absorption peaks at 520 cm^{-1}. The FESEM image showed the high degree of crystallinity of the synthesized CuO NPs in which surface morphology was found to be homogeneous and almost spherical in shape with the presence of capping agents that stabilize the CuO NPs. The thermal analyses were carried out using DSC and TGA techniques in which the synthesized CuO NPs were responded with temperature. The magnetic properties of prepared CuO NPs are examined by VSM which demonstrates that the synthesized CuO NPs showed slightly superparamagnetic characteristics with zero corrosivity and remanence magnetization property.

Acknowledgements The authors are grateful to the GARE (Grant of Advanced Research in Education) project, Ministry of Education, GoB for financing during the research work (Project/User ID: PS2019949).

References

1. Dr. Khan, M.K.A.: An Introduction to Solar Energy. Mizan Library, Dhaka, Bangladesh (2006)
2. Rai, G.D.: Solar Energy Utilization. Khanna Publishers, Delhi-110006, India (2004)
3. Gunawan, et al.: Energy storage system from galvanic cell using electrolyte from a plant as an alternative renewable energy. IOP Conf. Ser.: Mater. Sci. Eng. **509**, 012045 (2019)
4. Hamid, M.R.: Characterization of a battery cell fueled by bryophyllum pinnatum sap. Int. J. Sci. Eng. Res. **4**(3), 1–4 (2013). ISSN 2229-5518
5. Hamid, M.R., Yusuf, A., Wadud, A.M.A., Rahaman, M.M.: Design and performance test of a prototype of a 12 Volt DC battery fueled by bryophyllum pinnatum sap and improvement of its characteristics, department of electrical and electronic engineering, Ahsanullah university of science and technology, Dhaka, Bangladesh. Int. J. Electron. Electr. Eng. **4**(5), 398–402 (2016)
6. Elgammal, A., Sharaf, A.M.: Dynamic self adjusting FACTS-switched filter compensation, schemes for wind-smart grid interface systems Int. J. Renew. Energy Res. **2**(1), 103–111 (2016)
7. Nedaei, M.: Wind energy potential assessment in Chalus county in Iran. Int. J. Renew. Energy Res. **2**(2), 338–347 (2012)
8. Tzen, E., Papapetrou, M.: Promotion of renewable energy sources for water production through desalination. Desalination Water Treat. **39**(1–3), 302–307 (2012)
9. Ying, C.Y., Dayou, J.: Modelling of the electricity generation from living plants. J. Teknol. **78**(6), 29–33 (2016)
10. Chee, F.P., Chen, C.A., Chang, J.H.W., Choo, Y.Y., Dayou, J.: Data acquisition system for in situ monitoring of chemoelectrical potential in living plant fuel cells. J. Biophys. (2016)
11. Volkov, A.G., Foster, J.C., Jovanov, E., Markin, V.S.: Anisotropy and nonlinear properties of electrochemical circuits in leaves of Aloe vera L. Bioelectrochem. **81**(1), 4–9 (2011)
12. Choo, Y.Y., Dayou, J., Surugau, N.: Origin of weak electrical energy production from living-plants Int. J. Renew. Energy Res. **4**(1), 198–203 (2014)
13. Ghate, N.B., Hazra, B., Sarkar, R., Mandal, N.: vitro anticancer activity of Spondias pinnata bark on human lung and breast carcinoma. Cytotechnology **66**(2), 209–218 (2014)
14. Kamal, S., Akhter, R., Tithi, N.A., Wadud, M.A., Narjish, S.N., Shahriar, M., Bhuiyan, M.A.: Biological investigations of the leaf extract of Spondias pinnata. Int. J. Pharm. Sci. Rev. Res. **6**(8), 3351 (2015)
15. Bogdanski, S., Ashton, A., Fogg, A., Burns, D.T., Headridge, J., Sowerbutts, A., East, G., Bishop, E., Bark, L., Grime, J.: Research topics in analytical chemistry. Proc. Soc. Anal. Chem. **9**(9), 182–211 (1972)
16. Lau, O.W., Luk, S.F.: A survey on the composition of mineral water and identification of natural mineral water. Int. J. Food Sci. Technol. **37**(3), 309–317 (2002)
17. Khan, M.K.A., Paul, S., Rahman, M.S., Kundu, R.K., Hasan, M.M., Moniruzzaman, M., Mamun, M.A.: A study of performance analysis of PKL electricity generation parameters: (An experimental analysis on voltage regulation, capacity and energy efficiency of pathorkuchi leaf (PKL) electricity cell). In: Power India International Conference (PIICON), 2016 IEEE 7th, 2016, pp. 1–6. IEEE (2016)

18. Khan, M.K.A., Rahman, M.S., Das, T., Ahmed, M.N., Saha, K.N., Paul, S.: Investigation on parameters performance of Zn/Cu electrodes of PKL, AVL, tomato and lemon juice based electrochemical cells: a comparative study. In: Electrical Information and Communication Technology (EICT), 2015 3rd International Conference on, pp. 1–6. IEEE (2017)
19. Khan, K.A., Rahman, A., Rahman, M.S., Tahsin, A., Jubyer, K.M.: Paul, S.: Performance analysis of electrical parameters of PKL electricity (An experimental analysis on discharge rates, capacity & discharge time, pulse performance and cycle life & deep discharge of Pathor Kuchi Leaf (PKL) electricity cell).In: Innovative Smart Grid Technologies-Asia (ISGT-Asia), 2016 IEEE, pp. 540–544. IEEE (2016)
20. Khan, M.K.A., Paul, S., Rahman, M.S., Kundu, R.K., Hasan, M.M., Moniruzzaman, M., Al Mamun, M.:.A study of performance analysis of PKL electricity generation parameters: (An experimental analysis on voltage regulation, capacity and energy efficiency of pathor kuchi leaf (PKL) electricity cell). In: Power India International Conference (PIICON), 2016 IEEE 7th, pp. 1–6. IEEE (2016)
21. Paul, S., Khan, K.A., Islam, K.A., Islam, B., Reza, M.A.: Modeling of a Biomass Energy based (BPL) Generating Power Plant and its features in comparison with other generating Plants. IPCBEE **44** (2012). IACSIT Press, Singapore.https://doi.org/10.7763/IPCBEE.2012.V44.3
22. Khan, K.A., Paul, S.: A analytical study on electrochemistry for PKL (Pathor Kuchi Leaf) electricity generation system. In: Publication date 2013/5/21, Conference- Energytech, 2013 IEEE, pp. 1–6. Publisher, IEEE (2013)
23. Hasan, M., Khan, K.A.: ***Bryophyllum pinnatum*** leaf fueled cell: an alternate way of supplying electricity at the off-grid areas in Bangladesh. In: Proceedings of 4th International Conference on the Developments in Renewable Energy Technology [ICDRET 2016], p. 01, (2016). https://doi.org/10.1109/ICDRET. 7421522
24. Hasan, M., Khan, K.A., Mamun, M.A.: An estimation of the extractable electrical energy from bryophyllum pinnatum leaf. Am. Int. J. Res. Sci-Ence, Technol. Eng. Math. (AIJRSTEM) **01**(19), 100–106 (2017)
25. Khan, M.K.A., Rahman, M.S., Das, T., Ahmed, M.N., Saha, K.N., Paul, S.: Investigation on parameters performance of Zn/Cu electrodes of PKL, AVL, tomato and lemon juice based electrochemical cells: a comparative study. In: Electrical Information and Communication Technology (EICT), 2017 3rd International Conference on, pp. 1–6. IEEE (2017). https://doi.org/10.1109/EICT.2017.8275150, IEEE, Khulna, Bangladesh, Bangladesh, 7–9
26. Duffie, J.A., Beckman, W.A.: Solar Energy Thermal Process. John Wiley and Sons. Inc N.Y. USA
27. Sharma, B.K.: (1998) Electrochemistry, Krishna Prakashan Media (P) Ltd., 11, Shivaji Road, Meerut-250 001 (UP) India (1998)

Applications of PKL Electricity for Use in DC Instruments

Lovelu Hassan and K. A. Khan

1 Introduction

The voltaic cell is also known as a galvanic cell which is an electrochemical cell and where spontaneous reactions for electricity generation occur. Generally, the voltaic cell has two half cells with a salt bridge. Our designed and fabricated PKL electrochemical cell has no salt bridge but the basic principle is based on voltaic or galvanic cell [1, 2]. That is why we have called it modified voltaic or modified galvanic cell. Sometimes it is also called called voltaic or galvanic cell. The main difference between the voltaic/galvanic electrochemical cell and PKL electrochemical cell is that no large current for voltaic/galvanic cell whereas a large current can be generated by PKL electrochemical cell [3, 4]. A PKL half cell has an electrode within a PKL extract with Mn+ ions, where M is a metal stip and n+ is the number of ions [5]. There is a link for two half cells by a connecting wire from one electrode to the other. The salt bridge in a voltaic cell connects to the half cells [6, 7]. The electrons are transferred and the energy is released due to the spontaneous redox (reduction–oxidation) reactions. As a result, it is possible to use that energy if the flow of electrons occurred through the external circuit [8, 9]. Figure 1 shows a typical PKL electrochemical cell. In our innovated PKL electrochemical cell, oxidation occurs at anode and reduction occurs at cathode [10, 11]. In the traditional voltaic cell, electrons flow from anode to cathode and the charges in each beaker would not be balanced and then the flow of electrons would stop [12]. Generally, a salt bridge looks like a u-tube which contains the solution of the salt and it keeps the charges balanced [13]. In the solution of the beaker, anions move toward the anode and cations move toward the cathode [14]. In our innovated modified voltaic cell, electrons leave the anode toward the cathode through an external circuit and the cations dissolve into the

---

L. Hassan
Department of Physics, Jahangirnagar University, Savar, Dhaka, Bangladesh

K. A. Khan (✉)
Department of Physics, Jagannath University, Dhaka 1100, Bangladesh

A. Biswas et al. (eds.), *Microelectronics, Circuits and Systems*, Lecture Notes in Electrical Engineering 755, https://doi.org/10.1007/978-981-16-1570-2_18

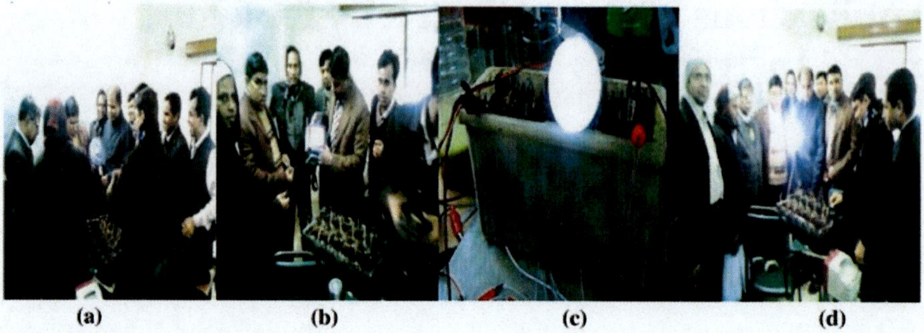

Fig. 1 Experimental setup for the DC appliances of PKL electricity. **a** DC table fan **b** DC TV **c** DC bulb **d** DC Energy bulb

PKL extract in each compartment. The PKL electricity generation device is based on Zn/Cu electrodes [15]. That is why when the electrons reached the Cu Plate, and cations in the Cu plate are attracted to the Cu plate [16]. The electrons are taken by the cations (Cu2+) and the neutral Cu metal is deposited on the Cu plate. The generated PKL electricity travels in a complete electrical circuit [17]. The spontaneous reaction of a PKL circuit always runs in that direction where the +ve electrochemical cell potential is produced. The main things for our complete PKL electric systems are obtained: the created cell potential of the PKL system, the direction of the flow of electrons, the role of anode (Zn plate) and cathode (Cu plate), the role of PKL extract, and the nature of ions, electrodes, and electrolytes [18].

1.1 Methodology

The different Specifications of Electrolyte is given below:

Quantity of PKL = 800 g = 0.8 kg, Quantity of water = 3200 ml = 3.2liter, and PKL: Water = 1:4, and the dimensions of the Zn electrode is length of the Zn electrode = 14.8 cm, width of the Zn Electrode = 3.8 cm, and the area of the Zn electrode = $(14.8 \times 3.8)\ cm^2 = 56.24\ cm^2$. The dimensions of each chamber is length of each chamber = 10.0 cm, width of each chamber = 3.5 cm, and the area of each chamber = $(10.0 \times 3.5)\ cm^2 = 35\ cm^2$.

It is shown in Fig. 1a the DC table fan with PKL electricity, in Fig. 1b about the DC TV with PKL electricity, in Fig. 1c regarding the DC LED bulb, and in Fig. 1d DC Energy bulb with PKL electricity. Overall Fig. 1 shows the experimental setup for the DC appliances of PKL electricity.

Electrodes: The zinc plate was used as an anode and the copper plate was used as a cathode. There are six chambers/compartments in a PKL module, which was made by an insulated battery box. There are one pair of electrodes (one Zinc plate and one copper plate)set up in each compartment. Then electrodes of each compartment are connected in a series combination. As a result, the voltage of the PKL module increases, and the current of the PKL module remains constant. Then after the DC load is connected between the ends of the PKL module. If the load needs high voltage, then more PKL module is connected in series connection to face the demand of the DC load voltage. The electrolyte prepared by PKL extract has been put into each compartment of the PKL module. After a chemical reaction, the voltage is generated in the PKL module which lights the DC appliances.

Electrolytes: PKL extract is mixed with water for getting electrolyte of the PKL electrochemical cell. This PKL extract has been filtered and then kept into each compartment of the PKL electrochemical cell. When the electrodes Zinc and Copper are immersed into the electrolyte, then the voltage and current are produced which lights the DC appliances.

Battery box: The battery box was made of an insulated material like plastics. Generally, it is available in the local market.

Connecting wires: The connecting wires are made of copper. It is also available in the market. It can be purchased from both local and international markets. Although it is also available in the local market, it can be purchased in kg or feet basis.

Multimeters: The voltage and current have been measured by a calibrated multimeters. It can be purchased from both local and international markets. Although it is also available in the local market.

PKL cultivation: The PKL is called a miracle leaf. Because it can grow everywhere. So that it can be harvested easily. After collecting the PKL, it has been blended by a blender or any other suitable crash machine. Then it is mixed with water. If the water is DI water, then it would better for better performance of electricity generation.

Direct Current (DC) and Alternating Current (AC): The current which flows in a unidirectional way is called Direct Current (DC). It follows the ohm's law, $V = IR$ and Power law, $P = VI$, where, V = voltage, I = current, and R = resistance. The current which changes its magnitude and polarity at regular intervals of time is called Alternating Current. The equation of ac voltage is: $v(t) = v_m \sin \omega t$ and $i(t) = i_m \sin \omega t$, where, $v(t)$ = ac voltage, v_m = maximum voltage, $i(t)$ = ac current, i_m = maximum current, and ω = angular frequency = $2\Pi f$. Here, f = Linear frequency. The ac power, $P = v(t)\, i(t) = (v_m{}^2/R) \sin^2 t$, where R = load resistance.

Inverter: It converts DC to AC. There are three types of inverters such as square wave inverter, modified square wave inverter or quasi inverter, and pure sine wave inverter.

1.2 Chemical Reactions

The difference between cathode and anode for a PKL electrochemical cell is given in Table 1. It is found that from Table 1 that Copper works as a cathode and Zinc works as an anode. It is also shown that the loss of electrons occurs at Zinc and gain of electrons occurs at Copper. The redox reaction occurs between Copper and Zinc electrodes.

Generation of Electric potential in a PKL System:

The potential difference between the Cu plate and Zn plate of a PKL cell is called the electric potential or electromotive force of a PKL electrochemical cell. Generally, it is denoted by E_{cell}. It is measured by the equation

$$1V = 1J/1C \tag{1}$$

(since, V = W/Q),

where V = potential, W = Potential Energy, and Q = charge. The pull or driving force on electrons of a PKL electrochemical cell is also called the electromotive force (emf). The cause of the electro motion in a PKL electrochemical cell is the electromotive force. The potential at the Zn plate may be called the oxidation potential and the potential at the Cu plate may be called the reduction potential. The electrode potential cannot be measured in isolation. The electrode potential of a PKL cell depends upon the chemical properties of the materials of the electrodes, concentration of the materials, temperature, and pressure.

Standard State Condition for PKL electrochemical cell:

The concentration of the solution = 1 mol/L, the pressure of gases = 1 atm, and temperature = 25°C or 298 K.

Standard PKL electrochemical Cell Potentials:

The PKL electrochemical cell potential at standard state conditions can be shown by the following equation:

$$\text{Ecell} = \text{Ered (cathode)} - \text{Ered (anode)} \tag{2}$$

Table 1 Difference between cathode and anode for a PKL electrochemical cell

Cathode (Copper electrode)	Anode (Zinc electrode)
Here electrons are consumed/gained	Here electrons are generated/lost
Here reduction occurs	Here oxidation occurs
Here positive(+) sign	Here negative(-) sign

Table 2 Difference between cathodic and anodic reactions for a PKL electrochemical cell [22].

Cathodic reactions	Anodic reactions
$Cu^{2+} + 2e^- = Cu$	$Cu\text{-}2e^- = Cu^{2+}$
$Zn^{2+} + 2e^- = Zn$	$Zn\text{-}2e^- = Zn^{2+}$
$Fe^{3+} + e^- = Fe^{2+}$	$Fe^{2+} – e^- = Fe^{3+}$
$H^+ + 2e^- = 2H = H2$	$2H – e^- = H^+$

where E°cell = Cell potential, E°red (cathode) = Cathode (Cu plate) potential or reduction potential, and E°red (anode) = Anode (Zn plate) potential or oxidation potential [19]. The intensive property of the PKL electrochemical cell can be found according to Eq. (1), and the potential energy per unit of charge is equal to the potential of the PKL electrochemical cell. For PKL electrochemical cell, E°red (cathode) = −0.76 V and E°red (anode) = +0.34 V. Therefore, from Eq. (2), we have E°cell = +0.34–(−0.76) = 1.10 V. The strongest oxidizers have the most +ve reduction potential and the strongest reducers have the most –ve oxidation potential. It is found that the PKL cell potential depends upon the difference of the potential of two electrodes. It is also shown that the reduction potential of the PKL cell potential does not depend upon the stoichiometric constant.

Summary of the standard electrode potential:

The electric voltage is created of a PKL electrochemical cell for two dissimilar electrode materials. This voltage has been measured for the potential energy per unit charge from the redox reaction to drive the total reaction activities. The PKL electrochemical cell has two half cells made by cathode (Cu plate) and anode (Zn plate) with PKL extract [20]. The important and interesting matter is that the created voltage of the PKL electrochemical cell is positive because the reaction is spontaneous.

1.3 Difference Between Cathodic and Anodic Reactions of a PKL Electrochemical Cell

It is shown in Table 2 that the Zn^{2+} is the product ion and the Cu^{2+}, Zn^{2+}, and H^+ are the reactant ions. In PKL electrochemical system, the concentration of the reactant ions should be higher than the concentration of the product ions [21].

2 Results and Discussion

The performance of PKL electrochemical cell for some parameters has been studied for two consecutive weeks which have been graphically presented.

The variation of open circuit (V_{OC}) with the variation of time duration for day-1is shown in Fig. 2. The V_{oc} indicates the average value of V_{oc1}, V_{oc2}, V_{oc3}, V_{oc4}, V_{oc5},

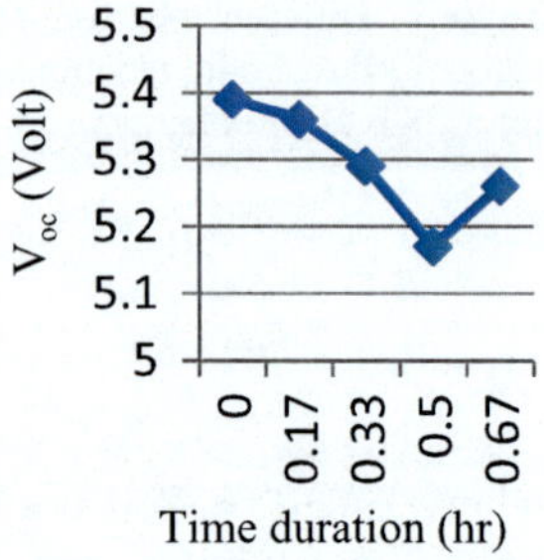

Fig. 2 V_{OC}–Time duration curve

and V_{oc6}. It is shown in Fig. 2 that the variation of open circuit voltage (V_{OC}) was from 5.17 V to 5.39 V for first day. Therefore, the open circuit voltage change was 0.22 V for first day. The variation of open circuit (V_{OC}) with the variation of time duration for day-2 is shown. The V_{oc} means the average value of V_{oc1}, V_{oc2}, V_{oc3}, V_{oc4}, V_{oc5}, and V_{oc6}. The variation of open circuit voltage from 5.05 V to 5.17 V for the second day is also shown in Fig. 3. It is mentioned that the difference between day-1 and day-2 is around one week. Therefore, the open circuit voltage change was 0.12 V for second day. Figure 4 shows the behavior of the variation of open circuit, V_{oc} for first day and second day together. The variation of open circuit V_{oc} for two different days is also found which has been recorded carefully and graphically shown. In Fig. 4, it is shown that the variation of open circuit (V_{OC}) was decreased by the PKL electrochemical cell after a week.

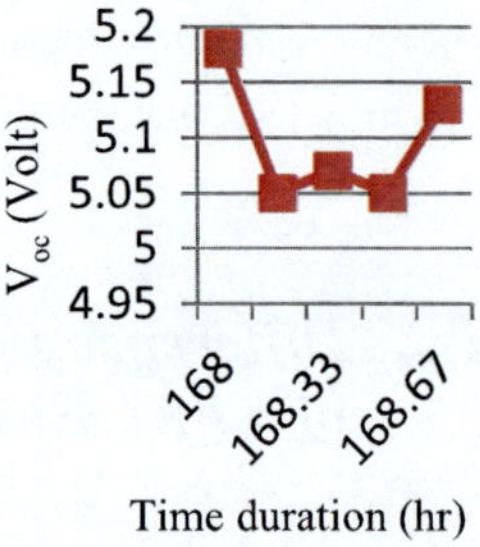

Fig. 3 V_{OC}–Time duration curve

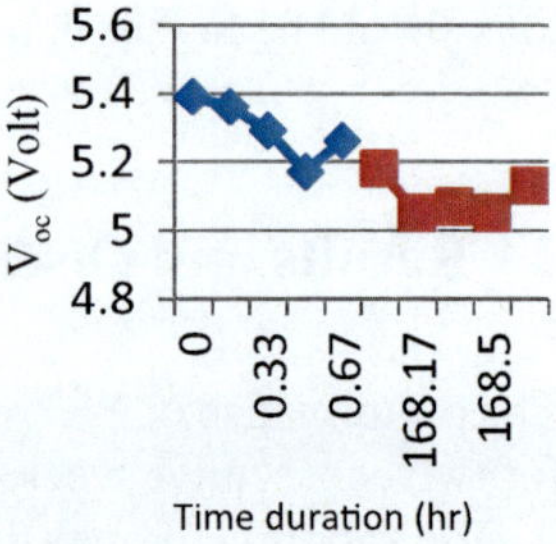

Fig. 4 V_{OC}–Time duration curve

The variation of short circuit current (I_{sc}) with the variation of time duration for day-1 is shown in Fig. 5. The I_{sc} indicates the average value of I_{sc1}, I_{sc2}, I_{sc3}, I_{sc4}, I_{sc5}, and I_{sc6}. It is shown in Fig. 5 that the variation of short circuit current (I_{sc}) was from 2.51 to 3.11 mA for first day. Therefore, the short circuit current (I_{sc}) change was 0.60 mA for first day. The variation of I_{sc} with the variation of time duration for day-2 is also shown. Here, the I_{sc} means the average value of I_{sc1}, I_{sc2}, I_{sc3}, I_{sc4}, I_{sc5}, and I_{sc6}. The variation of short circuit current (I_{sc}) from 2.50 to 3.40 mA for the second day is also shown in Fig. 6. It is mentioned that the difference between day-1 and day-2 is around one week. Therefore, the short circuit current change was 0.90 mA for the second day. Figure 6 shows the behavior of the variation of I_{sc} for the first[st] day and second day together. Thevariation of short circuit current (I_{sc}) for two different days is also found which has been recorded carefully and graphically shown. In Fig. 7, it is shown that the short circuit current (I_{sc}) was regained by the PKL electrochemical cell after a week.

Fig. 5 I_{SC}–Time duration curve

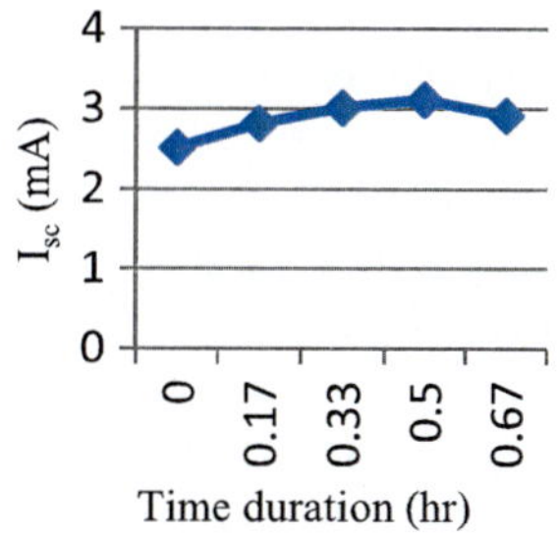

Fig. 6 I_{SC}–Time duration

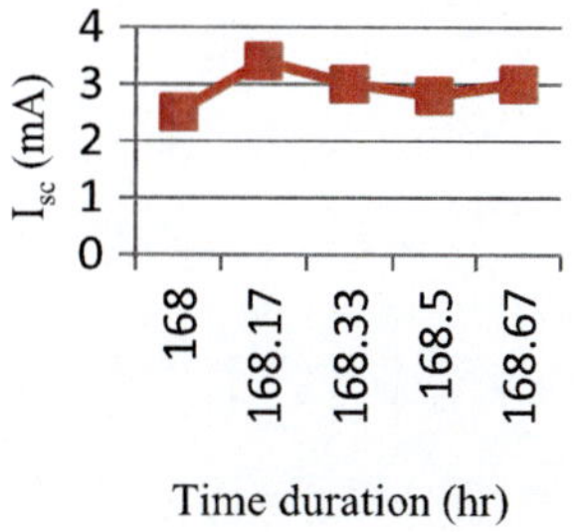

Fig. 7 I_{sc}–Time duration curve

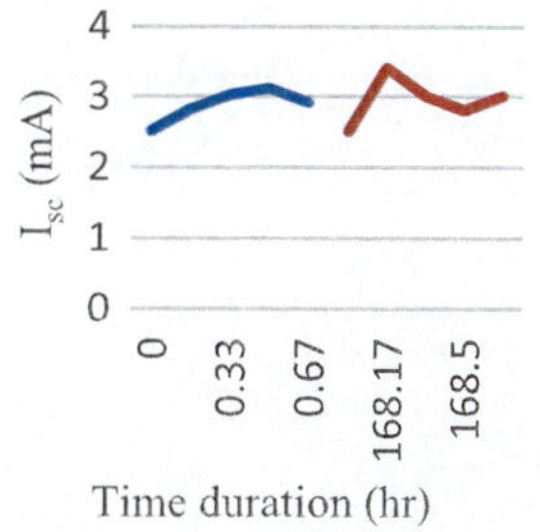

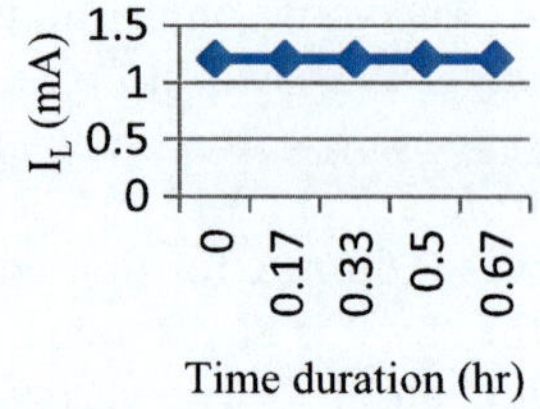

Fig. 8 I_L–Time duration curve

The variation of load current (I_L) with the variation of time duration for day-1 is shown in Fig. 8. The I_L indicates the load current. It is shown in Fig. 7 that the variation of load current (I_L) was from 1.21 mA for the first day. Therefore, the load current (I_L) was constant and there was no change of load current for the first day. The variation of I_L with the variation of time duration for day-2 is also shown. It is also shown in Fig. 9 that the variation of load current (I_L) was from 1.10 to 1.20 mA for the second day. It is mentioned that the difference between day-1 and day-2 is around one week. Therefore, the load current change was 0.10 mA for the second day. Figure 10 shows the behavior of the variation of I_L for the first day and second day together. The variation of short load current (I_L) for two different days is also found which has been recorded carefully and graphically shown. In Fig. 10, it is shown that the load current (I_L) was almost constant by the PKL electrochemical cell after a week.

The variation of maximum power (P_{max}) with the variation of time duration for day-1 is shown in Fig. 11. The P_{max} indicates the maximum power. It is shown in Fig. 11 that the variation of maximum power (P_{max}) was from 13.53 to 16.08 W for first day. Therefore, the maximum power (P_{max}) increased exponentially and the change of the maximum power was 2.55 W for first[t] day. The variation of P_{max} with

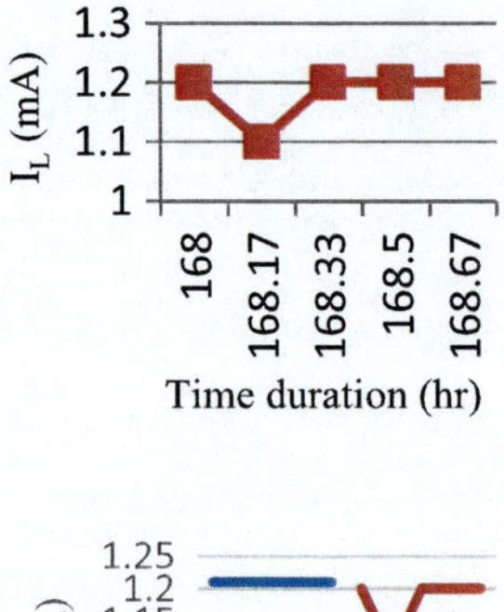

Fig. 9 I_L–Time duration curve

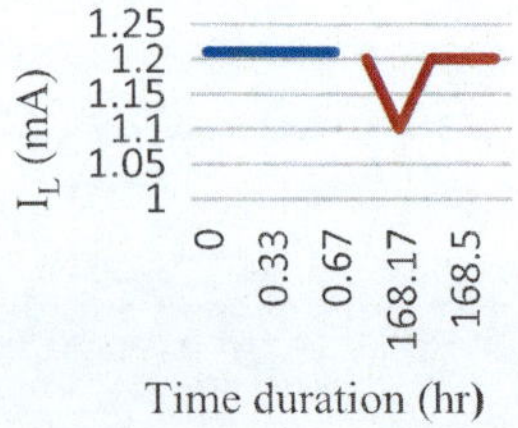

Fig. 10 I_L–Time duration curve

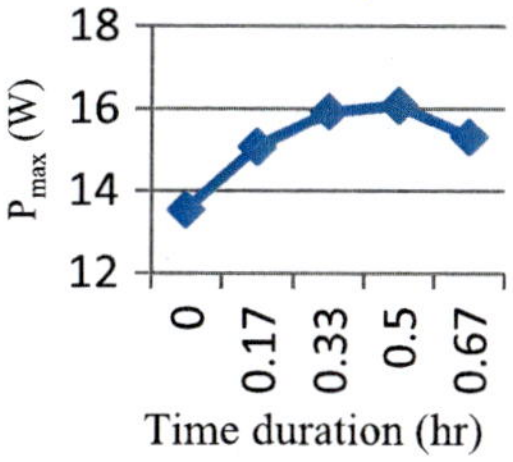

Fig. 11 P_{max}–Time duration curve

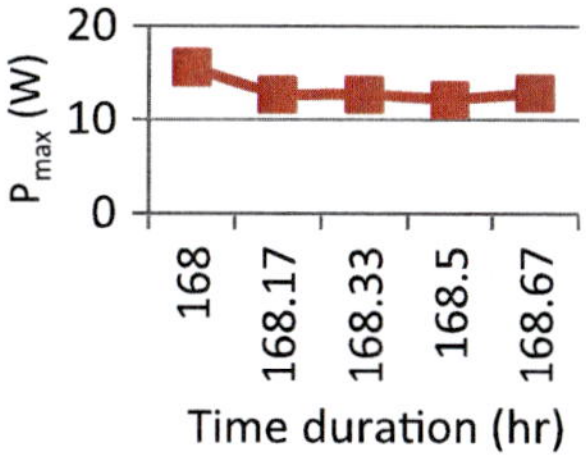

Fig. 12 P_{max}–Time duration curve

the variation of time duration for day-2 is also shown in Fig. 11. It is also shown in Fig. 12 that the variation of maximum power (P_{max}) was from 12.12 to 15.54 W for the second day. It is mentioned that the difference between day-1 and day-2 is around one week. Therefore, the maximum power change was 3.42 W and it was decreased about exponentially for the second day. Figure 13 shows the behavior of the variation of P_{max} for the first day and second day together. The variation of maximum power (P_{max}) for two different days is also found which has been recorded carefully and graphically shown. In Fig. 13, it is shown that the maximum power (P_{max}) was almost opposite for the PKL electrochemical cell after a week. So it can be concluded here that the maximum power of the PKL electrochemical cell has decreased a little bit after a week.

The variation of load power (P_L) with the variation of time duration for day-1 is shown in Fig. 14. The P_L indicates the load power. It is shown in Fig. 14 that the variation of load power (P_L) was from 2.58 to 2.62 W for first day. Therefore, the load power (P_L) was changed exponentially from 2.59 to 2.62 W firstly and then decreased also linearly up to 2.61 W, and the total change of the load power was 0.14 W for first day. The variation of P_L with the variation of time duration for day-2 is also shown

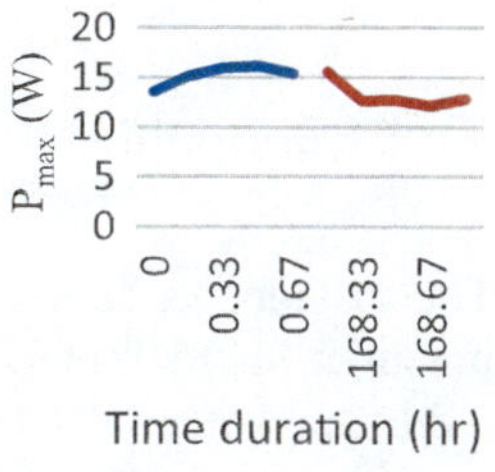

Fig. 13 P_{max}–Time duration curve

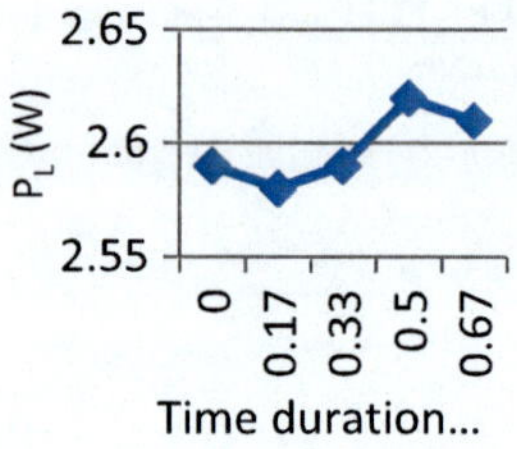

Fig. 14 P_L–Time duration curve

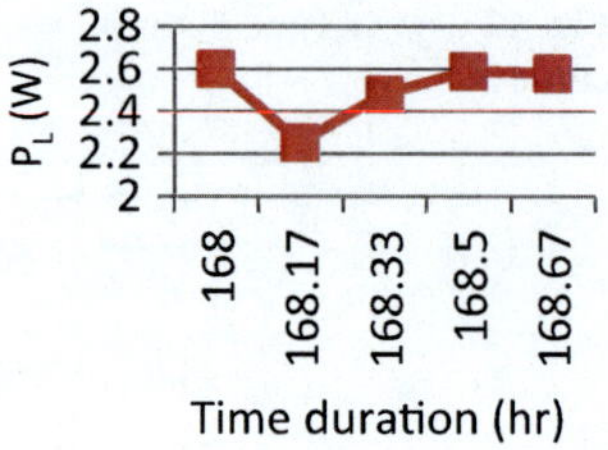

Fig. 15 P_L–Time duration curve

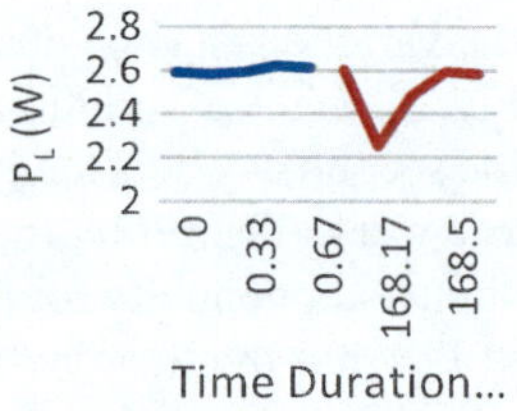

Fig. 16 P_L–Time duration curve

in Fig. 15. It is also shown in Fig. 15 that the variation of load power (P_L) was from 2.25 to 2.60 W for the second day. It is mentioned that the difference between day-1 and day-2 is around one week. Therefore, the load power change was 0.35 W and it was decreased firstly about linearly and then increased exponentially for the second day. Figure 16 shows the behavior of the variation of P_L for first day and second day together. The variation of load power (P_L) for two different days is also found which has been recorded carefully and graphically shown. In Fig. 16, it is shown that the load power (P_L) was almost opposite for the PKL electrochemical cell after a week. So it can be concluded here that the load power of the PKL electrochemical cell has decreased a little bit after a week.

3 Conclusions

The stronger oxidizing agent and the weaker reducing agent make a positive electro potential for PKL electrochemical system. It is in oxidized form for more negative reduction potential which makes the weaker oxidizing agent and it is in reduced form for the stronger reducing agent. The positive value of cell potential indicates the overall reaction is spontaneous. The electricity is produced due to the voltage difference between two Zn and Cu electrodes for PKL electrochemical cell. This voltage difference is also the difference between two individual voltages of Zn and Cu electrodes with respect to the PKL extract electrolyte. The overall voltage of PKL electrochemical cell can be measured properly but there is no simple way to properly measure the voltage between electrode–electrolyte till today. The electric voltage of the PKL electrochemical cell depends upon concentration, temperature, and pressure.

Acknowledgements The authors are grateful to the GARE (Grant of Advanced Research in Education) project, Ministry of Education, GoB for financing during the research work (Project/User ID: PS2019949).

References

1. Hamid, M.R.: Characterization of a battery cell fueled by *Bryophyllum Pinnatum* sap. Int. J. Sci. Eng. Res. **4**(3), 1–4 (2013) ISSN 2229-5518
2. Hamid, M.R., Yusuf, A., Wadud, A.M.A., Rahaman, M.M.: Design and performance test of a prototype of a 12 Volt DC battery fueled by *bryophyllum pinnatum* sap and improvement of its characteristics, Department of electrical and electronic engineering, Ahsanullah university of science and technology, Dhaka, Bangladesh. International Journal of Electronics and Electrical Engineering, **4**(5), 398–402 (2016)
3. Khan, K.A., Wadud, M.A., Obaydullah, A.K.M., Mamun, M.A.: PKL (Bryophyllum Pinnatum) electricity for practical utilization. IJARIIE-ISSN(O)-2395-4396, **4**(1), 957–966 (2018)
4. Ohiduzzaman, M., Khatun, R., Reza, S., Khan, K.A., Akter, S., Uddin, M.F., Ahasan, M.M.: Study of exposure rates from various nuclear medicine scan at INMAS. Dhaka. IJARIIE **5**(3), 208–218 (2019)
5. Khan, K.A., Rasel, S.R.: Development of a new theory for PKL electricity using Zn/Cu electrodes: per pair per volt. IJARIIE **5**(3), 1243–1253 (2019)
6. Khan, K.A., Abu Salek, M.: A study on research development and demonstration of renewable energy technologies. IJARIIE **5**(4), 113–125 (2019)
7. Khan, K.A., Uddin, M.N., Nazrul Islam, Md., Mondol, N., Ferdous, Md.: A study on some other likely renewable sources for developing countries. IJARIIE **5**(4), 126–134 (2019)
8. Khan, K.A., Zian Reza, S.M.: The situation of renewable energy policy and planning in developing countries. IJARIIE **5**(4), 557–565 (2019)
9. Khan, K.A., Rasel, S.R., Reza, S.M.Z., Yesmin, F.: Electricity from living PKL tree, Published in the open access book. In: Hermoso-Orzáez, M.J. (eds.) Energy Efficiency and Sustainability in Outdoor Lighting A Bet for the Future. London, UK (2019)
10. Gunawan, et al.: Energy storage system from galvanic cell using electrolyte from a plant as an alternative renewable energy. IOP Conf. Ser.: Mater. Sci. Eng. **509**, 012045 (2019)
11. Nedaei, M.: Wind energy potential assessment in Chalus county in Iran Int. J. Renew. Energy Res. **2**(2), 338–347 (2012)
12. Tzen, E., Papapetrou, M.: Promotion of renewable energy sources for water production through desalination. Desalination Water Treat. **39**(1–3), 302–307 (2012)
13. Ying, C.Y., Dayou, J.: Modelling of the electricity generation from living plants. J. Teknol. **78**(6), 29–33 (2016)
14. Chee, F.P., Chen, C.A., Chang, J.H.W., Choo, Y.Y., Dayou, J.: Data acquisition system for in situ monitoring of chemoelectrical potential in living plant fuel cells. J. Biophys. (2016)

15. Volkov, A.G., Foster, J.C., Jovanov, E., Markin, V.S.: Anisotropy and nonlinear properties of electrochemical circuits in leaves of Aloe vera L. Bioelectrochemistry **81**(1), 4–9 (2011)
16. Choo, Y.Y., Dayou, J., Surugau, N.: Origin of weak electrical energy production from living-plants Int. J. Renew. Energy Res. **4**(1), 198–203 (2014)
17. Khan, K.A., Bhuyan, M.S., Mamun, M.A., Ibrahim, M., Hasan, L., Wadud, M.A.: Organic electricity from Zn/Cu-PKL electrochemical cell. In: Mandal, J.K. et al. (eds.) Contemporary Advances in Innovative and Applicable Information Technology, Advances in Intelligent Systems and Computing, ©Springer Nature Singapore Pvt. Ltd., 2018, Vol. 812, Chap. 9, pp. 75–90 (2018)
18. Khan, K.A., Rasel, S.R., Zian Reza, S.M., Yesmin, F.: Energy Efficiency and Sustainability in Outdoor Lighting—A Bet for the Future, Energy Efficiency and Sustainable Lighting—a Bet for the Future, Manuel Jesús Hermoso-Orzáez and Alfonso Gago-Calderón, IntechOpen (2020).https://doi.org/10.5772/intechopen.89413
19. Khan, K.A., Yesmin, F., Md. Wadud, A., Obaydullah, A.K.M.: Performance of PKL electricity for use in television. Accepted as a book chapter NAROSA publisher (2019)
20. Rab, M.N.F., Khan, K.A., Rasel, S.R., Ali, M.H., Hassan, L., Salek, M.A., Reza, S.Z., Ohiduzzaman, M.: Voltage cultivation from fresh leaves of air plant. In: Climbing Spinach, Mint, Spinach and Indian Pennywort for Practical Utilization. Energy Systems, Drives andAutomations, Proceedings of ESDA 2019, Springer Singapore, Lecture Notes in Electrical Engineering, eBook ISBN: 978-981-15-5089-8. https://doi.org/10.1007/978-981-15-5089-8. Hardcover ISBN: 978-981-15-5088-1, Series ISSN: 1876-1100, Vol. 664, pp. 150–160 (2020)
21. Khan, K.A., Rasel, S.R., Reza, S.Z., Saime, M.A., Alam, N., Salek, A., Hasan, M.: Solar medical sterilizer using pressure cooker for rural off-grid areas. In: Energy Systems, Drives andAutomations, Proceedings of ESDA 2019 , Springer Singapore, Lecture Notes in Electrical Engineering, eBook ISBN: 978-981-15-5089-8. https://doi.org/10.1007/978-981-15-5089-8. Hardcover ISBN: 978-981-15-5088-1, Series ISSN: 1876-1100, Vol. 664, pp. 258–269 (2020)
22. Khan, K.A., Saime, M.A., Ali, M.H., Reza, S.Z., Alam, N., Hossain, M.A., Rab, M.N.F., Islam, S.: A study on PKL electrochemical cell for three different conditions. In: Energy Systems, Drives andAutomations, Proceedings of ESDA 2019 , Springer Singapore, Lecture Notes in Electrical Engineering, eBook ISBN: 978-981-15-5089-8. https://doi.org/10.1007/978-981-15-5089-8, Hardcover ISBN: 978-981-15-5088-1, Series ISSN: 1876-1100, Vol. 664, pp. 374–386 (2020)

Fast Charging Method for Wireless Power Transfer Enabled Devices

Shikhar Jain and Aman Sharan

1 Introduction

Wireless power transmission additionally termed as wireless charging is an engineering science that permits the relay of non-particulate radiation energy from the source to an electrical load through the air gap in the absence of the use of interlinking cables or wires. This cable-free novelty is accentuating in an in-depth sort of implementation from a low-power mobile charger to a high-power electrical automobile understanding of its convenient operation and simplicity of execution. The wireless power transfer phenomenon operates in the following process. Initially, a voltage source is imperative to operate the AC to DC rectifier. Wireless power transfer requires high frequency for its effective operation. Commercially a frequency of 50 Hz is used in various parts of the world, also there are countries like USA which use 60 Hz to operate their various electrical machineries. However, these frequencies are too low to drive the charging circuit. Using a combination of rectifier and inverter, this AC supply is converted to DC, which is followed by raising the voltage of DC and converting the DC back to high-frequency AC using full-wave inverters. This high-frequency AC is input to the transmitter coil in the shape of a loop which produces a magnetic flux around it. An alternating current is induced at the receiver coil which is placed at a certain distance from the transmitter coil due to mutual induction between the two coils. The terminals of the receiver coil are then connected to the rectifier coils so that the alternating current is changed into direct current via the rectifier. The output terminals of the rectifier are then connected to the rechargeable battery socket that can be used to recharge the batteries (Fig. 1).

S. Jain (✉) · A. Sharan
Department of Electrical Engineering, Delhi Technological University, New Delhi, India

A. Biswas et al. (eds.), *Microelectronics, Circuits and Systems*, Lecture Notes in Electrical Engineering 755, https://doi.org/10.1007/978-981-16-1570-2_19

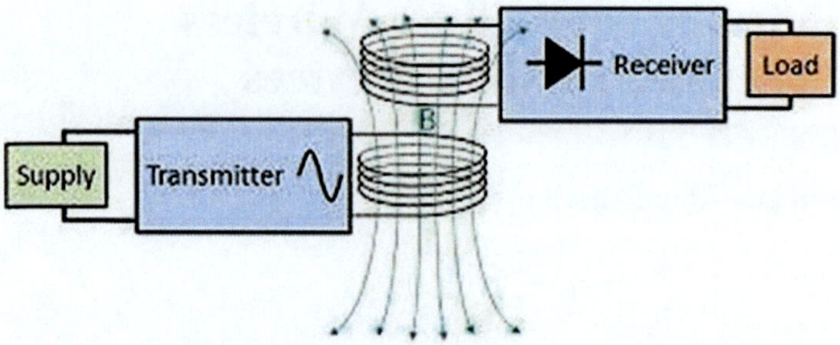

Fig. 1 Working of a wireless power transfer system

2 Classification of Wireless Power Transfer Techniques

The Wireless Charging Technologies can be broadly classified into three main categories:

1 **Non-Radiative Charging**: Also popularly known as near-field charging, Non-radiative charging is the technique in which the power is transferred over a short distance by a magnetic field using inductive coils. These can be subdivided into two types:

   1.1 **Inductive coupling**: Similarities exist between the Inductive coupling and the transformer coupling as both primarily focus on the transfer of power based on electromagnetic induction.

   1.2 **Magnetic Coupling**: It is similar to the inductive coupling but the difference is that the coupled coils are connected to the capacitor and a condition of resonance is achieved which permits loose alignment and has higher efficiency in comparison to the inductive coupling.

2 **Radiative Charging**: Also known as far-field charging has the advantage of charging at a distance of typically a few meters to a few kilometers. The mode of transmission of power is mostly RF, but requires large transmitting and receiving antennas.

3 **Acoustic Charging**: In this technological advancement, the ultrasonic which are sound waves are used to enable power transmission.

3 Applications of Wpt

Wireless Power Transfer finds its application ranging from compact electronic devices up to the more heavier and complex electric vehicles. Firstly, WPT could replace the traditional wired charging systems that are used in the present times. Instead of charging electronic devices such as mobile phones via a power cord, wireless power helps to replenish the battery without the actual use of any connecting cables. WPT has major research potential in Electric vehicle charging. As the demand for EVs is on the continuous rise, charging these vehicles efficiently and economically still poses a greater challenge for the scientists. The reliability and demand of driving such vehicles for a better and sustainable future solely depend upon the charging time, availability of charging stations, and the ease of driving such vehicles over the traditionally fuel-fired ones. WPT along with some advanced power electronics circuit provides one such alternative to this problem.

4 Wireless Charging System

4.1 Center Tapped Transformer

The construction of a center tap transformer is such that it provides dual independent secondary output voltages, V1 and V2, to work along a general connection. The output of the present type of transformer composition results in a 2-phase, 3-wire supply. The connection point along the transformer winding to select the suitable number of turns on the secondary winding is the exact center point that provides a typical connection which induces two equal yet opposite secondary voltages. As the tapping at the center of the transformer is grounded, the two output voltages V1 and V2 are positive and negative with respect to the earthing point and also with a 180° phase shift between each other, respectively.

1. Vertical mount type
2. Supply Voltage: 220 V AC at 50 Hz
3. Vout: 24, 12 V, or 0 V
4. Iout: 5A (Fig. 2).

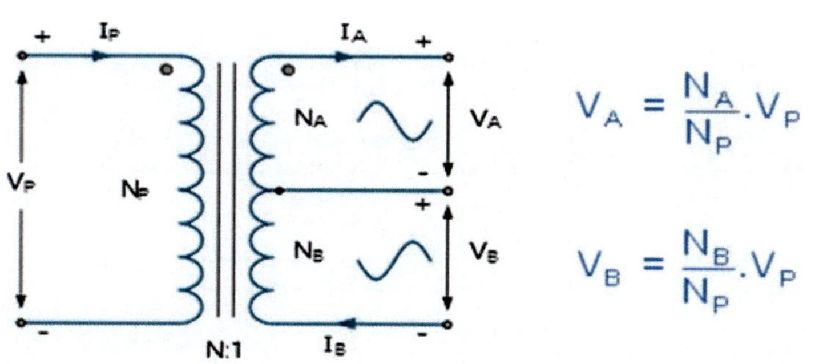

Fig. 2 Stepping down of 220 V input voltage into 12 V output voltage using a center tap transformer

4.2 AC–DC Rectifier

Rectifier circuits are those circuits that aid in the conversion of alternating current (AC) into direct current (DC). The 12 V alternating output voltage is tapped from the transformer and is input to the rectifier circuit. The circuit converts the 12 V AC input voltage to a constant DC output voltage of the same magnitude. In the first half-cycle (positive) of the alternating voltage, the first terminal is positive, center tap is at a null potential, and the second terminal will be at a negative value. This leads to a forward bias of diode D1 and causes current to pass through diode D1. Simultaneously, diode D2 is reverse biased and does not allow current to flow through it. When the polarity is changed, i.e., during the negative half-cycle of the input voltage, the second terminal will become positive with respect to the first terminal and the center tap. This results in a forward bias of diode D2 and current starts flowing through it. During this time, diode D1 is reverse biased and does not allow the passage of current through it (Fig. 3).

4.3 DC–AC Inverter

The world today operates on 50 or 60 Hz. But during wireless power transfer, the main problem is of efficiency, and to achieve effective and maximum transfer of power, higher frequency is required. To achieve this high-frequency AC, the 12 V DC output of the rectifier is fed to an inverter circuit comprising of MOSFETs, inductors, and capacitors. This circuit helps to convert the low-frequency AC input current into high-frequency AC which is further provided to the transmitting coil to generate a magnetic field around it (Fig. 4).

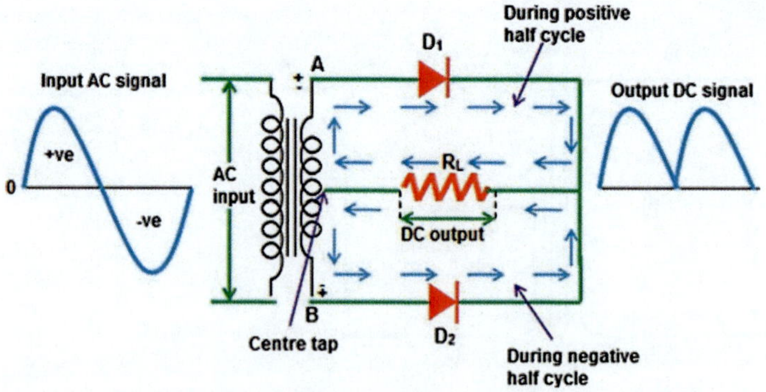

Fig. 3 Rectification of 12 V AC from the transformer into DC

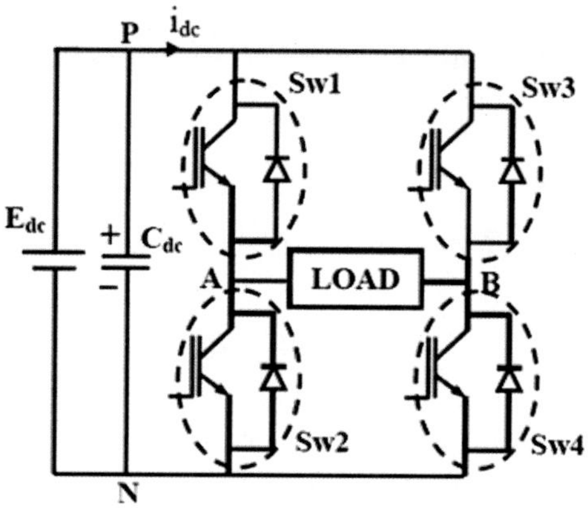

Fig. 4 A full-wave inverter circuit comprising of MOSFETs, inductors, and capacitors

Table 1 Specification of the coils used

Parameter	Transmission coil	Receiving coil
Material	Copper	Copper
Thickness	26 SWG	26 SWG
Number of turns	150	180

4.4 Transmitting and Receiving Circuits

The Transmitter coil (Tx) and Receiver coil (Rx) are inductive coil. It is observed that maximum power transfer takes place when both the coils overlap each other. In this case, there is maximum flux linkage between both the Tx and the Rx coils. As the distance between the coils is increased, the magnetic field generated by the Tx coil does not overlap with the Rx coil. As a result, the power transferred decreases and ultimately becomes zero at a distance of a maximum of 20 cm in our case.

Ideally, the number of turns of the primary and the secondary coils should be the same. But due to the presence of leakage flux and losses, the secondary coil has a greater number of turns so as to counter these losses. We also use ferrite core so as to increase the flux linkages (Table 1).

4.5 LM317EMP

The LM217-N is a device consisting of adaptable three-pin positive voltage regulators which are effective in providing a surplus of 1.4A over a 1.3–35 V output domain and a broad temperature range. It is extraordinarily effortless to utilize and calls for only a pair of peripheral resistors to specify secondary voltage. The line regulation of the instrument is better than other line regulators. Moreover, load regulation is superior to typical load regulators. The LM317EMP offers complete overload protection such as thermal overload safety, safe area protection, and current limit. Entire overload protection circuitry is always completely utilitarian although the adjusting terminal is disengaged.

4.6 *The Auto-Cut Arrangement*

The auto-cut arrangement constitutes of a Zener diode, Relay, and transistor switch.

(1) **Transistor switch**—BD139 is an NPN transistor. which has a gain value of 40 to 160. This value determines the amplification capacity of the transistor. The maximum passage of the current permitted through the collector pin is 1.5A. To bias a transistor and for its effective operation, a base current (IB) needs to be provided to the base terminal. This current should be limited to 1/10th of the collector current and the voltage across the base emitter pin should be a maximum of 5 V.

(2) **Zener diode**—The role of the diode is to start conducting as soon as the voltage is above the prespecified value so as to allow the transistor to conduct.

(3) **Relay**—1-Channel 12 V Relay Module is a device used for the closing of the switch when connected to the battery for charging. The relay's primary utility in the given circuit is to open the circuit when the voltage across the battery rises above 12 V. As soon as the Zener diode starts conducting, the transistor is triggered. The transistor acting as a switch is short circuited and the current starts flowing in the magnetizing coil of the relay. The coil attracts the conductor and opens the circuit due to electromagnetic attraction (Fig. 5).

5 Working

Following is the way in which the circuit works.

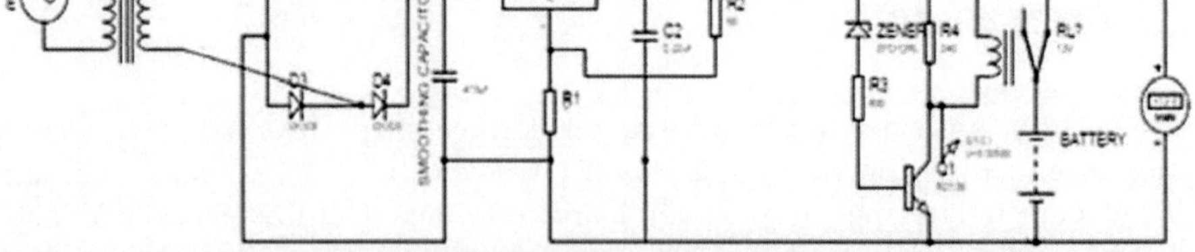

Fig. 5 Circuit for fast charging mechanism

1. The supply is given to the single-phase center tap transformer (220 V/12–0-12 V). The transformer steps down the voltage to 12 V.
2. The transformer is then connected to a closed-loop rectifier that is responsible for the conversion of the alternating current to direct current. Due to the maximum efficiency of 80%, the output voltage from the rectifier is around 8.5 V.
3. The rectified DC current then flows through the inverter which converts it into AC and is then electrically coupled to the transmitting coil. The output voltage from the inverter is 6.94 V.
4. The transmitting coil (150 turns) is magnetically coupled with the receiving coil (180 turns).
5. Ignoring the losses, the receiving coil voltage is 9.18 V.
6. The receiver coil is connected to the bridge rectifier having a rectification efficiency of 80%, hence the output 6.5 V.
7. The current is then fed to the input terminal of the voltage regulator.
8. The output from the voltage regulator is calculated as Voutput = 1.25 V(1 + R1/R2) + Iadjust(R1)
9. A diode is connected so as to promote the flow of current from the regulator to the battery and not vice-versa.
10. When the value of the voltage across the diode is more than the prescribed value, the Zener diode starts conducting, and the base terminal of the transistor receives current.
11. When the value of the current in the base terminal of the transistor (base current) exceeds the specified value, the transistor which is working as a switch is triggered and the switch is closed.
12. When the switch is closed, the relay gets operated and the battery is isolated leading to auto-cut-off.

6 Calculations

A 220/12 V transformer is supplied a voltage of 220 V.

RMS Voltage from the transformer = 12 V.

It is fed into the full-wave rectifier, where the AC is converted into DC. The output of full-wave rectifier is given by

$$=> \text{Vdc1} = \text{Vrms}/\sqrt{2}$$

$$=> \text{Vdc1} = 0.70712$$

$$=> \text{Vdc1} = \mathbf{8.5V}$$

The full-wave rectifier is then connected to the inverter to convert the DC again into AC due to the requirement of high frequency. The fundamental RMS output voltage from the inverter

$$=> \mathrm{E1} = \sqrt{2/3} \times \mathrm{Vdc1}$$

$$=> \mathrm{E1} = \sqrt{2/3} \times 8.5$$

$$=> \mathrm{E1} = \mathbf{6.94V}$$

Due to complete overlap between the windings,

$$=> \frac{N1}{N2} = \frac{E1}{E2}$$

$$=> \frac{150}{180} = \frac{6.94}{E2}$$

where N1 and N2 are the number of turns 150 and 180, respectively on both the primary and the receiver side

$$=> \mathrm{E2} = \mathbf{9.18V}$$

The output of the full-wave rectifier on the receiver side is

$$=> \mathrm{Vdc2} = \mathrm{E2}/\sqrt{2}$$

$$=> \mathrm{Vdc2} = 0.7079.18$$

$$=> \mathrm{Vdc2} = \mathbf{6.5V}$$

Voltage through the voltage regulator

$$=> \mathrm{I\ adj} = 100\ \text{(From Data Sheet)}$$

$$=> \mathrm{Vout} = \mathrm{V\ ref}\ (1 + \mathrm{R1/R2}) + \mathrm{I\ adj}\ (\mathrm{R1})\ \text{(From Data Sheet)}$$

$$=> \mathrm{V\ ref} = 1.25\mathrm{V}\ \text{(From Data Sheet)}$$

$$\mathrm{R1} = 1.8\mathrm{K}\Omega \mathrm{R2} = 220\Omega)$$

$$=> \text{Vout} = 1.25(1) + 100\,(1800) = 11.65\text{V}$$

7 Conclusion

This paper presented the design and implementation of a wireless power transfer system including fast charging technology. To achieve high output power and power transfer efficiency, an inverter, rectifier, transmitter, and receiver coil and regulator were optimally designed. Using wireless power technology eliminates the need of using cord connections to replenish portable electronic devices. The main problems that are being faced today in wireless power transfer are of slow charging and low efficiency. A fast-charging mechanism has been proposed in this paper. It was observed that maximum flux linkage takes place when the coils overlap each other. As the distance between the coils is increased, the flux linkage decreases which hampers the power transfer and output efficiency. Therefore, inductive wireless charging is seen to be an efficient method for charging compact electronic devices, which can be kept near to the charging circuit for effective operation, minimum losses, and maximum efficiency.

Design and Simulation of Bi-metallic RF MEMS Switch for Fast Switching Time

K. Srinivasa Rao, Sk. Shoukath Vali, K. Girija Sravani, P. Ashok Kumar, and Koushik Guha

1 Introduction

MEMS switches are the combination of both mechanical and electrical structures. The mechanical configuration switches are designed as Fixed-Fixed, and cantilever beams and electrical configurations are based on series or shunt type such as metal to metal contact, capacitive type of RF MEMS switches [1]. Capacitive MEMS switches are preferred in RF application due to their high isolation and low insertion losses at high frequencies. RF MEMS switches having tremendous advantages over commercial switches like Gallium Arsenide (GaAs), Filled Effected Transistor (FET) and photo-injected diode (PIN Diode). The utilization of RF MEMS switches in communication [2], space application fields exponentially improved due to low power consumption, low noise, low cost, low weight, high linearity and lower intermodulation distortion [3, 4]. Surface micromachining process is the suitable method to fabricate the RF MEMS switches having a thin metallic beam called bridge over an electrode, and beam can allow and block the electronic transmission by mechanical moment of the bridge. To perform switching operation, electrostatic force will be generated by applying voltage beam and lower electrode [5]. The RF MEMS switch suffers from low switching time and high actuation voltages. A novel RF MEMS capacitive switch with step structure reduces the gap between beam and dielectric, a helical shaped spring for unique spring constant produces low actuation voltage of 2.2 V, S11 < −11.5 db, S21 > −0.85 db in upstate at 1–40 GHz and high isolation −71 db in down state condition at 30.5 GHz [6]. Superior changes in insertion losses

K. Srinivasa Rao (✉) · Sk. Shoukath Vali · K. Girija Sravani · P. Ashok Kumar
Department of ECE, MEMS Research Center, Koneru Lakshmaiah Educational Foundation (Deemed to Be University), Guntur 522502, Andhra Pradesh, India
e-mail: drksrao@kluniversity.in

K. Girija Sravani · K. Guha
Department of ECE, National MEMS Design Centre, National Institute of Technology (NIT), Silchar, Assam, India

A. Biswas et al. (eds.), *Microelectronics, Circuits and Systems*, Lecture Notes in Electrical Engineering 755, https://doi.org/10.1007/978-981-16-1570-2_20

and return loses for a shunt and series RF MEMS switch will be acquired by using asymmetric structure [6, 7]. Non-uniform RF MEMS shunt type switch is designed to attain less spring constant, small pull-in voltage and also high isolation [8–10].

This paper presents the comparison of electromechanical characteristics of normal beam and bimorph beam of RF MEMS capacitive switch. The remaining part of the paper follows: the proposed design is explained in Sect. 2, in Sect. 3, the results carried out using FEM tool are presented and followed by the conclusion in Sect. 4.

2 Proposed Design

The proposed switch consists of a silicon substrate having a length, width and height of 700 μm, 410 μm and 150 μm, respectively. Over a substrate, 0.5 μm-thick silicon dioxide layer is placed. Coplanar waveguide (CPW) is positioned on an oxide layer, to transmit RF signal. The width of the signal line is 55 μm and the separation from the ground with is 120 μm. Over a signal line, dielectric material of SI_3N_4 is kept with a length and width of 350*410 μm. The performance characteristics of RF MEMS switch depending on the membrane dimensions and material. The design of the membrane has some specifications like low pull-in voltage and low spring constant [11]. Low spring constant will be achieved by using serpentine flexure-type meanders. Fixed-Fixed membrane is placed over the dielectric with the help of anchors, the gap between the beam and dielectric is 2 μm. The thickness of the beam is one of the key parameters to control the pull-in voltage of the switch, if the thickness of the beam is very high, it required more actuation voltage to pull down the beam towards dialectic to close the switch. When the switch is in a close position, no signal will transmit from the input to the output port. All dimensions of the proposed switch are explained in Fig. 1.

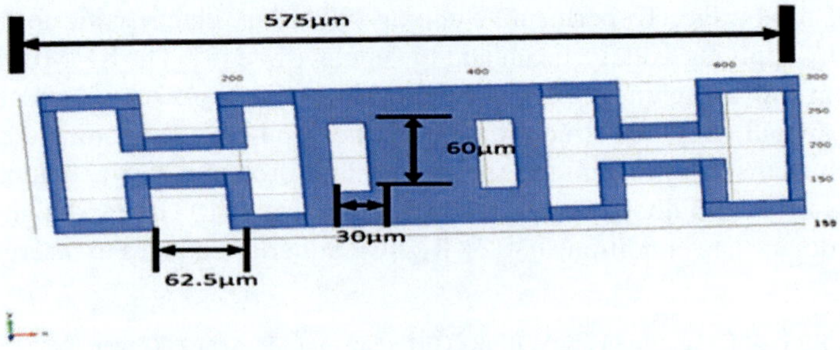

Fig. 1 Cross-sectional view of the proposed switch

2.1 Proposed Bimorph RF MEMS Switch

The design of the beam includes holes and meanders, two holes are inclusive on top of the beam with the dimensions 100*30 μm to reduce the residual stress and air damping. Serpentine flexure type meanders are used to pull down the total spring constant. The spring constant value of meanders is 5.16 N/m. The dimensions of beam with meanders are shown in Fig. 2. The design structure of the normal Fixed-Fixed beam is similar to the bi-metallic beam structure, instead of tacking single beam two beams are replacing with same dimensions. Selection of materials for bi-metallic beam are polymers and materials to reduce the mass of the beam [12, 13]. The proposed bimorph RF MEMS shunt switch is shown in Fig. 3 and the specific dimensions of both normal beam and bi-metallic beam are shown in Tables 1 and 2 respectively.

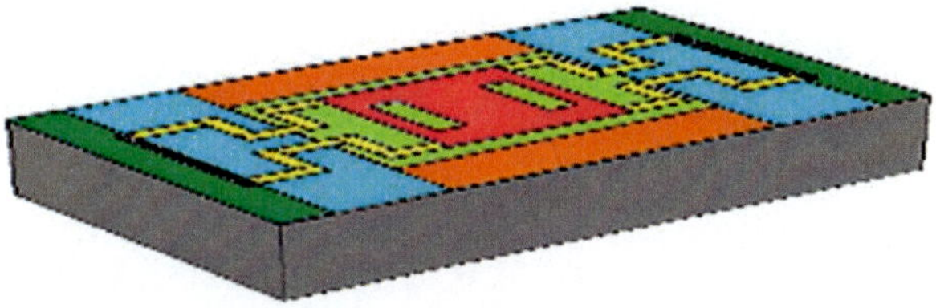

Fig. 2 Normal beam RF MEMS shunt switch

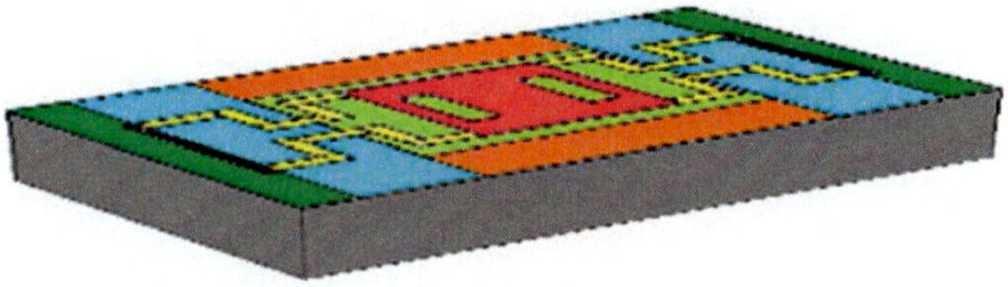

Fig. 3 Bimorph beam RF MEMS shunt switch

Table 1 Shunt beam switch specifications

S. No	Parameters	Length (μm)	Width (μm)	Depth (μm)	Material
1	Substrate	700	410	150	Silicon
2	CPW (G/S/G)	55	350	1	Gold
3	Beam	200	200	1	Gold
4	Holes	100	30	1	Gold
5	Dielectric layer	350	250	1	Si_3N_4

Table 2 Bimorph switch specifications

S. No.	Parameters	Length (μm)	Width (μm)	Depth (μm)	Material
1	Substrate	700	410	150	Silicon
2	CPW (G/S/G)	55	350	1	Gold
3	Beam	200	200	0.5 + 0.5	Gold, aluminum
4	Holes	100	30	1	Gold
5	Dielectric layer	350	250	1	Si_3N_4

3 Results and Discussion

3.1 Pull-in Voltage

The pull-in voltage of the desired switch is the voltage required to bring the beam to two-thirds of the distance, which can be obtained from Eq. 1 [14, 15]

$$V_p = \sqrt{\frac{8kg_0^3}{27\varepsilon_0 A}} \tag{1}$$

where k is the spring constant, ε_0 is the permittivity of free space, i.e. $\varepsilon_0 = 8.85 \times 10^{-12}$ F/m and overlapping area is A, g is gap between beam and dielectric. Spring constant and distance between the CPW and beam are mostly affecting the pull-in voltage of RF MEMS switch. Spring constant calculations per both normal and bi-metallic beams are similar, same type meander structures are used in both types of switches. The stiffness of the beam [16] can be obtained by

$$K = \frac{\mathrm{Ewt}^3}{\mathrm{l}^3} \tag{2}$$

where a single meander structure consists of five identical beams and each meander having a spring constant of

$$K_m = \frac{K}{5} \tag{3}$$

where w—beam width, l—beam length, t—beam thickness, E—Young's modulus.

The spring stiffness of the beam is obtained from Eq. 2 as 6.5 N/m. The spring constant of each meander is obtained from Eq. 3 as 1.296 N/m. The desired switch has four different meanders, which are used to parallel suspend the beam. Thus, the switch overall spring constant is four times the meander spring constant and is obtained as 5.16 N/m. Thus, the switch pull-in voltage is achieved by replacing the k value and the go value in Eq. 1 and is technically obtained as 4.6 V. The proposed switches using normal beam and bi-metallic beam is simulated using FEM tool in

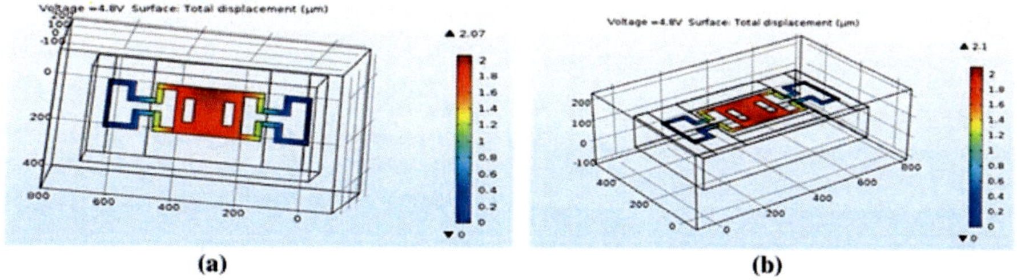

Fig. 4 **a** Pull-in voltage of normal shunt beam is 4.6 V, **b** pull-in voltage of bimorph beam is 4.6 V

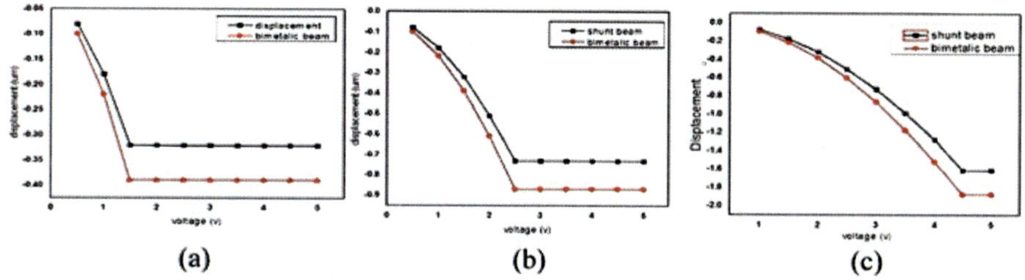

Fig. 5 Pull-in voltage RF MEMS switch at **a** 1 μm, **b** 1 μm, **c** 1 μm gaps for beam thickness 1 μm

electromechanics physics environment. When a voltage of 4.8 V is applied on the beam, the beam deflects all through the gap vertically downwards and falls on the dielectric layer. The normal beam switch and bi-metallic switch exhibit same pull-in voltage as the parameters involved in obtaining pull-in voltage are not affected by bi-metallic beam combination [13, 17]. The pull-in voltages of normal shunt beam and bimorph beam are shown in Fig. 4.

The pull-in voltage of RF MEMS switches at 1 μm thickness of the beam is represented in Fig. 5, the variation with respect to gap between beam and dielectric such as 1, 1.5 and 2 μm is analyzed by simulations. The minimum pull-in voltage has occurred at 1 μm gap and this low gap will enhance the capacitance of the switch [18, 19].

The pull-in voltage of RF MEMS switches at 0.5 μm thickness of the beam is represented in Fig. 6, varying with respect to different gaps between beam and dielectric such as 1, 1.5 and 2 μm. The minimum pull-in voltage is occurred at 1 μm

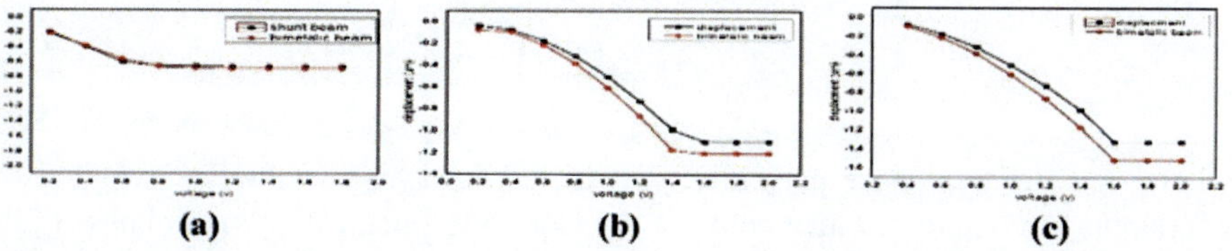

Fig. 6 Pull-in voltage RF MEMS switch **a**1 μm, **b** 1 μm, **c** 1 μm gaps with beam thickness 1 μm

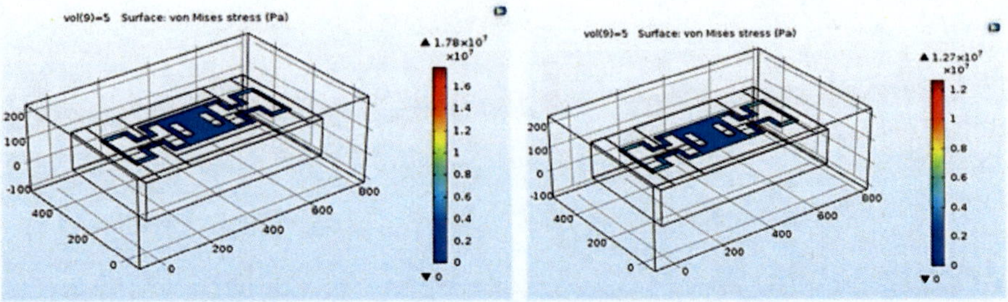

(a) (b)

Fig. 7 **a** Stress development in normal beam at 5 V, **b** stress development in bi-metallic beam at 5 V

gap, low gap and small thickness of the beam increase switch displacement due to high electrostatic forces [20].

From Figs. 4 and 5, the minimum gap between the beam and dielectric is 2 μm. The stress development in the membrane is not uniform in normal beam and bi-metallic beam due to the difference in the density of the bi-metallic beam. The normal beam switch shows high stress on the meanders, which is 17.8 MPa, whereas the bi-metallic beam shows low stress of 12.7 MPa. This is due to the decrease in mass of the bi-metallic beam. The stress development of normal shunt beam and bimorph beam are shown in Fig. 7.

3.2 *Capacitance Analysis*

Variation of capacitance with voltage applied between electrodes is plotted by electrostatic analysis. Good dielectric materials generally have a relative permittivity between 3.4 and 13. In the switch discussed, dielectric material is taken as silicon nitride, which has a dielectric constant of 7.5. The up capacitance of the switch can be assessed by Eq. 4.

$$C_u = \frac{\varepsilon_0 A}{g + \frac{t_d}{\varepsilon_r}} \tag{4}$$

The down state capacitance is given by

$$C_d = \frac{\varepsilon_0 \varepsilon_r A}{t_d} \tag{5}$$

where ε_0 is the free space permittivity, ε_r dielectric layer permittivity, A, g are the overlapping area and distance between the electrodes, respectively, thickness of the dielectric layer is t_d. Capacitance ratio is the ratio between upstate capacitance and the down state capacitance. The capacitance ratio is given by [18]

Fig. 8 Up state capacitance

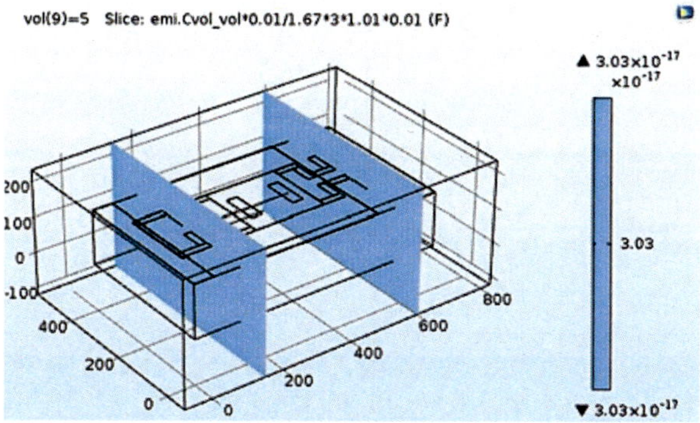

Fig. 9 Down state capacitance

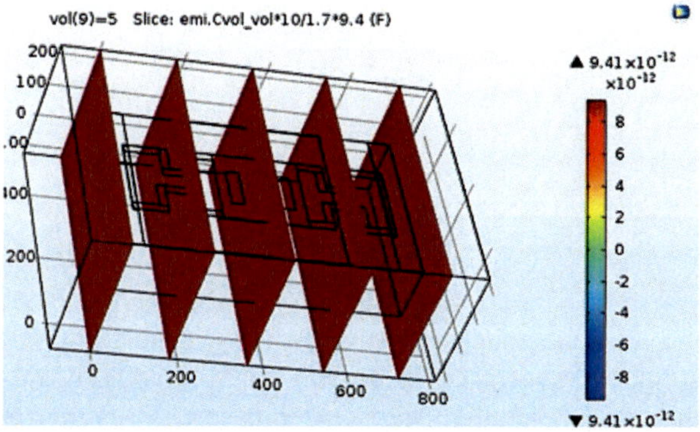

$$C_r = C_d / C_u \tag{6}$$

The down state and upstate capacitance is calculated from the above equations as 303.5 fF and 9 pF. The capacitance ratio of 29.7 is attained from the switch proposed. The capacitance of proposed RF MEMS switches depends on the surface area of beam; therefore, the bi-metallic beam combination does not affect the capacitance development of switch. Hence, two switches will have the same up and down state capacitances. The proposed switch is simulated using FEM tool in electrostatic environment and obtained the simulated results of 303.5 fF in upstate condition and 9 pF in down state condition. The simulated capacitance ratio of the switch is 30.3. The simulated upstate and downstate capacitance of both normal shunt beam and bimorph beam are similar, shown in Figs. 8 and 9 respectively.

The upstate capacitance with varying gaps 3, 2.5, 2, 1.5, 1 μm and dielectric thickness from 0.3 to 1 μm are represented in Fig. 10.

The dependency of capacitance over thickness of the dielectric layer and the air gap between the beam and dielectric are analyzed by plotting the graph as shown in Fig. 10. The switch shows high capacitance values at $g = 2$ μm; $t_b = 1$ μm; $t_d = 1$ μm.

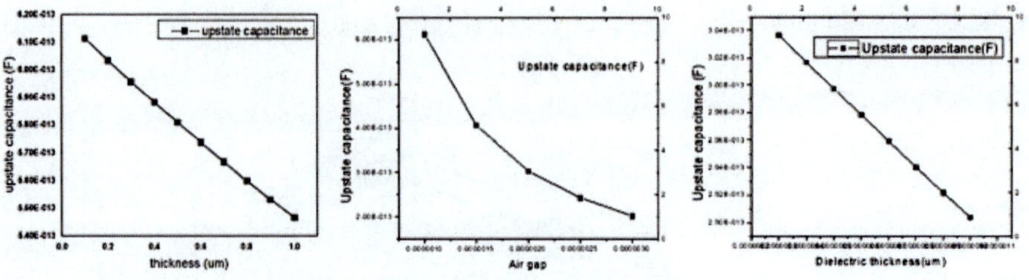

Fig. 10 Upstate capacitance variations by parametric analysis

3.3 Switching Time Analysis

The applied voltage is responsible for the changes in the switching time. Greater the applied voltage, stronger the electrostatic forces acting on the beam to displace it. Switching speed defines the device performance. The switching time is given by Eq. 9 [21].

$$T_s = \frac{3.67V_p}{V_s\omega_0} \tag{9}$$

where V_s, Vp are the actuation voltage and the pull-in voltage ($V_s = 1.4V_p$), and the angular resonant frequency is ω_0. The switching obtained for normal beam-structured switch is 8.6 μs and for bi-metallic beam switch is 7.2 μs. This shows that the bi-metallic switch exhibits fast switching mechanism than normal beam-structured switch. The comparison of both normal shunt type switch and proposed bi-metallic shunt switch is shown in Table 3. The switching time of both normal shunt beam and bimorph beam are shown in Fig. 11.

Table 3 Switch specifications

Sl. No.	Parameter	Normal shunt switch	Proposed bi-metallic shunt switch	Observation
1	Pull-in voltage	4.6 V	4.6 V	No change due same spring stiffness
2	Capacitance ratio	30.3	30.3	No change due to same overlapping area
3	Resonance frequency	19.58 kHz	23.4 kHz	Increased due to increase in mass of the beam
4	Quality factor	2.2	1.5	<2 (desired)
5	Switching time	8.6 μs	7.2 μs	Fast switching is achieved

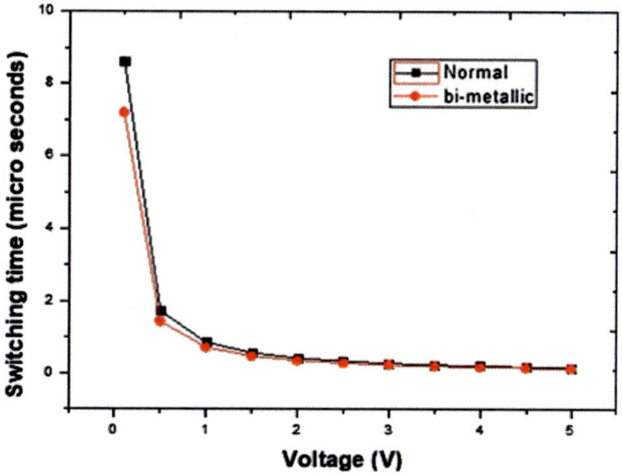

Fig. 11 Switching time graph of the switches

3.4 Electromagnetic Analysis

The S-parameters are observed by simulating switch using HFSS software at both onstate and offstate of switches. The insertion losses and return losses are measured in onstage, that is upstate position of the proposed switch (Figs. 12 and 13, respectively [7, 19]).

The switch shows less than 1 dB of insertion loss up to 16 GHz and shows good return losses up to 12 GHz. The switch is designed for 8–12 GHz for X-band applications. The isolation of proposed bi-metallic switch is shown in Fig. 14.

The switch has high isolation of −48.7 dB at 20 GHz and good isolation of −25.6 dB at 8–12 GHz. This shows that the switch is effectively working at 8–12 GHz frequency band for X-band applications [22].

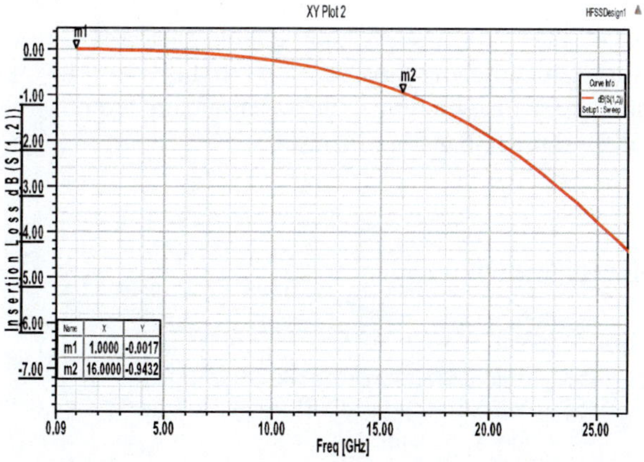

Fig. 12 Insertion loss at upstate

Fig. 13 Return loss at upstate

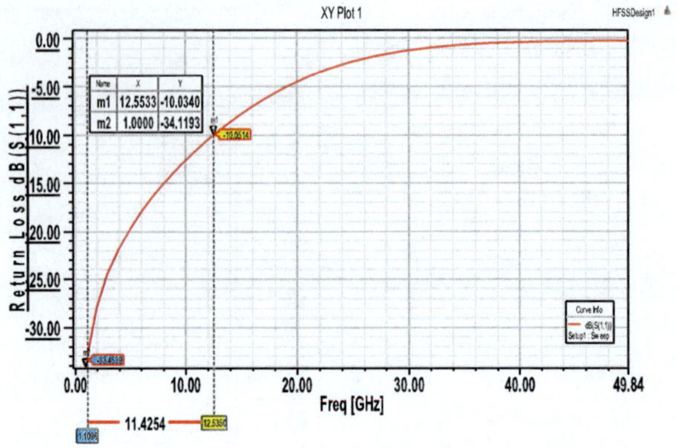

Fig. 14 Isolation at down state

4 Conclusions

The proposed design of normal beam and bi-metallic beam switches are analyzed for electromechanical and electromagnetic characteristics. The proposed bi-metallic beam structure shows low pull-in voltage of 4.8 V with high resonant frequency of 23.42 kHz and serves a good quality factor of 1.5. The bi-metallic switch shows a very fast switching time of 7.2 μs and is useful for many RF applications with very less delay in switching. The S-parameters are observed over 8–12 GHz and can be utilized for X-band applications.

References

1. Sravani, K.G., Prathyusha, D., Rao, K.S., Kumar, P.A., Lakshmi, G.S., Chand, C.G., Guha, K.: Design and performance analysis of low pull-in voltage of dimple type capacitive RF MEMS shunt switch for Ka-band. IEEE Access **7**, 44471–44488 (2019)
2. Chakraborty, A., Gupta, B., Sarkar, B.K.: Design, fabrication and characterization of miniature RF MEMS switched capacitor-based phase shifter. J. Microelectron. **45**, 1093–1102 (2014)
3. Rebeiz, G.M.: RF MEMS Theory, Design, and Technology, 1st edn. Wiley, New York (2003)

4. Srinivasa Rao, K., et al.: Design, modeling and analysis of perforated RF MEMS capacitive shunt switch. IEEE Access **7**, 74869–74878 (2019)
5. Ansari, H.R., Khosroabadi, S.: Design and simulation of a novel RF MEMS shunt capacitive switch with a unique spring for Ka-band application. Microsyst. Technol. https://doi.org/10.1007/s00542-018-3989-9
6. Srinivasa Rao, K., Lakshmi Narayana, T., Guha, K., Girija Sravani, K.: Fabrication and Characterization of Capacitive RF MEMS Perforated Switch, pp. 1–1. IEEE Access (2018). https://doi.org/10.1109/access.2018.2883353
7. GirijaSravani, K., Lakshmi Narayana, T., Srinivasa Rao, K.: A micro level electrostatically actuated cantilever and metal contact based series RF MEMS switch. Cogent Eng. (2017). https://doi.org/10.1080/23311916.2017.13233.67
8. Sravani, K.G., Guha, K., Rao, K.S.: Analysis on selection of beam material for novel step structured RF-MEMS switch used for satellite communication applications. Trans. Electr. Electron. Mater. **19**(6), 467–474 (2018)
9. Brown, E.R.: RF-MEMS switches for reconfigurable integrated circuits. IEEE Trans. Microw. Theory Tech. **46**(11) (1998)
10. Lakshmi Narayana, T., Girija Sravani, K., Srinivasa Rao, K.: Design and analysis of CPW based shunt capacitive RF MEMS switch. Cogent Eng. **4**(1) (2017)
11. Srinivasa Rao, K.: Design and performance analysis of uniform meandered structured RF MEMS capacitive shunt switchlong with perforations. Microsyst. Technol. **24**(2), 901–908 (2018)
12. Molaei, S., Ganji, B.A.: Design and simulation of a novel RF MEMS shunt capacitive switch with low actuation voltage and high isolation. Microsyst. Technol. **23**(6), 1907–1912 (2017)
13. Girija Sravani, K., Guha, K., Srinivasa Rao, K.: A modified proposed capacitance model for step structure capacitive RF MEMS switch by incorporating fringing field effects. Int. J. Electron. 1–22 (2020)
14. Nguyen, C.T.C.: Microelectromechanical devices for wireless communications. In: The 11th Annual International Workshop on Micro Electromechanical Systems, pp. 1–7, Heidelberg, Germany (1998)
15. Girija Sravani, K., Sudheer Babu, G., Ramesh, M., Prudhvi Raj, V., Prathyusha, D., Srinivasa Rao, K.: Beam and dielectric materials impact on improvement of the performance of a novel capacitive shunt RF MEMS switch. Int. J. Numer. Model.: Electron. Netw. Devices Fields **32**(5), e2598 (2019)
16. Jang, Y-H., et al.: High isolation RF MEMS contact switch in V-and W-bands using two directional motions. Electron. Lett. **46**(2), 153–155 (2010). In: Dielectric 3rd International Conference on Materials Processing and Characterization (ICMPC 2014)
17. Girija Sravani, K., Guha, K., Srinivasa Rao, K., Elsinawi, A.: Design of a novel structure capacitive RF MEMS switch to improve performance parameters. IET Circuits Devices Syst. **13**(7), 1093–1101 (2019)
18. Angria, M., Sundaram, G.M., Rangra, K.: Performance improvement of a reconfigurable series-shunt switch via asymmetric structure-based RF-MEMS capacitive switch. Int. J. Adv. Eng. Sci. Appl. Math. **7**(4), 198–203 (2015)
19. Iannacci, J., Huhn, M., Tschoban, C., Pötter, H.: RF-MEMS technology for 5G: series and shunt attenuator modules demonstrated up to 110 GHz. IEEE Electron. Device Lett. **37**(10), 1336–1339 (2016)
20. Sravani, K.G., Guha, K., Rao, K.S.: Analysis of a novel RF MEMS switch using different meander techniques. Microsyst. Technol. **26**(5), 1625–1635 (2020)
21. Schwartz, R.N., Wu, M., Hsu, T.Y., Schmitz, A.E., Loo, R.Y., Schaffner, J.H., Tangonan, G.L.: U.S. Patent No. 7,053,737. U.S. Patent and Trademark Office, Washington, DC (2006)
22. Ravirala, A.K., Bethapudi, L.K., Kommareddy, J., Thommandru, B.S., Jasti, S., Gorantla, P.R., Puli, A., Girija Sravani, K., Srinivasa Rao, K.: Design and performance analysis of uniform meander structured RF MEMS capacitive shunt switch along with perforations. Microsyst. Technol. **24**(2), 901–908 (2018). https://doi.org/10.1007/s00542-017-3403-z

Reconfigurable Antennas for RFID/GPS/WiMAX/WLAN Applications Using RF MEMS Switches

Lakshmi Narayana Thalluri, K. Srinivasa Rao, G. Venkata Hari Prasad, S. S. Kiran, Koushik Guha, Appala Raju Kanakala, and P. Bose Babu

1 Introduction

The demand for innovative wireless communication applications is drastically increasing day by day [1, 2]. To fulfill this demand, there is a necessity for novel and optimized communication devices like reconfigurable antennas, phase shifters, and RF isolators with low power consumption. Especially in 5G technology, there is a significant demand for reconfigurable antennas with multi-operating frequencies and low interferences [3, 4]. The researchers are investigated with intense curiosity to propose the optimized solution and to design a reconfigurable antenna with best performance. In this era, MEMS technology is shown great potential in the design of communication devices with low power consumption and high linearity.

This is the major motivated factor to design reconfigurable antennas by using RF MEMS switches other than traditional solid-state electronic switches (PIN diode and FET transistor) [5–7]. MEMS technology has recently received significant attention due to its ability to enhance the performance of RF switches. Despite the best

L. N. Thalluri (✉) · A. R. Kanakala · P. Bose Babu
Department of ECE, Andhra Loyola Institute of Engineering and Technology, Vijayawada, AP, India

K. Srinivasa Rao
Department of ECE, MEMS Research Center, Koneru Lakshmaiah Educational Foundation, Guntur, AP, India

G. Venkata Hari Prasad
Department of ECE, CMRCET, Kandlakoya, Hyderabad, Telangana, India
e-mail: drgvenkatahariprasad@cmrcet.org

S. S. Kiran
Department of ECE, Lendi Institute of Engineering and Technology, Vizianagaram, AP, India

K. Guha
Department of ECE, National MEMS Design Center, National Institute of Technology, Silchar, Silchar, Assam, India

A. Biswas et al. (eds.), *Microelectronics, Circuits and Systems*, Lecture Notes in Electrical Engineering 755, https://doi.org/10.1007/978-981-16-1570-2_21

Table 1 Wireless applications based on frequency bands [14–16]

Frequency range (GHz)	Application
0.9–5.2	RFID
1.2–1.6	GPS
1.85–1.9	GSM
2.39–2.4 and 5.2–5.6	WLAN
3.4–3.69	WiMAX
3.7–4.2 and 5.1–6.25 and 6.7–7.1	Few satellites

performance, decreasing the pull-in voltage, increasing the reliability, and improving the switching time are significant research challenges in RF MEMS switches. Improving the capacitance ratio and achieving the best RF performance are the potential challenges in capacitive RF MEMS switches [8–10]. RF MEMS switch-based reconfigurable antennas are offering the best performance when compared with structure-based reconfigurable antennas [11–13].

In this paper, we have presented an investigation and design analysis on the performance of reconfigurable antennas using shunt capacitive RF MEMS switches. These configurations aim to provide the following bands: GPS bands (1.2–1.6 GHz), WLAN bands (2.39–2.4 and 5.2–5.6 GHz), RFID (0.9–5.2 GHz), some satellite bands (3.7–4.2, 5.1–6.25, and 6.7–7.1 GHz), and GSM bands (1.85–1.9 GHz) (Table 1).

2 Reconfigurable Antenna

In this paper, we have designed a reconfigurable microstrip patch antenna using shunt capacitive RF MEMS switches. The antenna is aimed to serve multiple applications like RFID, WLAN, GPS, and WiMAX applications.

2.1 RF MEMS Switches

The shunt capacitive RF MEMS switches are used as the basic switching elements used for making the antenna as reconfigurable. The switch is designed with serpentine structure membrane of gold (Au) material of 1 μm thickness (Table 2; Figs. 1, 2 and 3).

The membrane is perforated with 5 μm × 5 μm size square shape holes. Electrostatic actuation is used to deform the membrane. The RF MEMS switch designed using COMSOL tool is offering good performance i.e., the required actuation voltage is 4.5 V, the switching time of the switch is 45 μs. We have done a wide range of analyses on the serpentine membrane in terms of max/min actuation voltages points, max/min displacement points, isolating surface for electric potential, displacement direction, and contour for electric potential, principal stress points.

Table 2 RF MEMS switch materials and dimensions

Parameter		Material	Dimension
Serpentine membrane thickness		Gold (Au)	1 μm
Perforation		–	5 μm × 5 μm
Dielectric		AlN	0.5 μm
Meander 10 μm uniform width 	1	Gold (Au)	20 μm
	2		60 μm
	3		70 μm
	4		80 μm
	5		30 μm
Air gap		–	1.5 μm

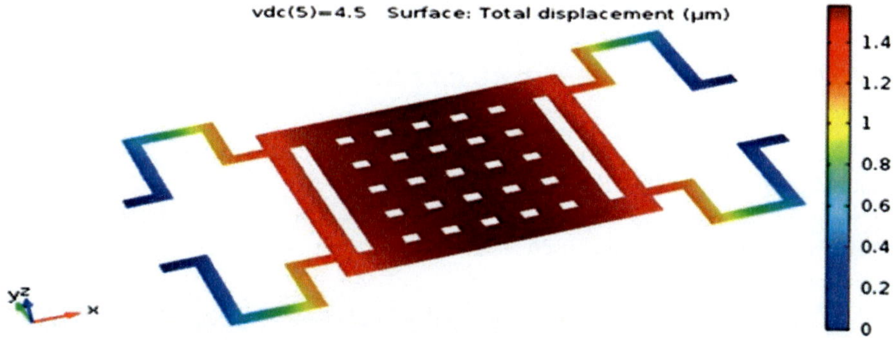

Fig. 1 Serpentine membrane electrostatic actuation

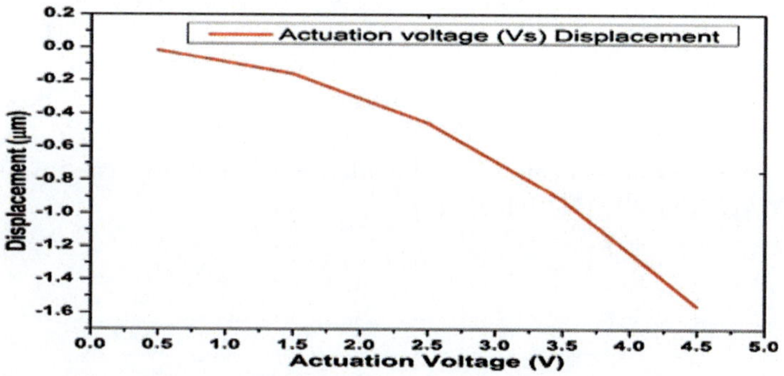

Fig. 2 Actuation voltage (Vs) displacement

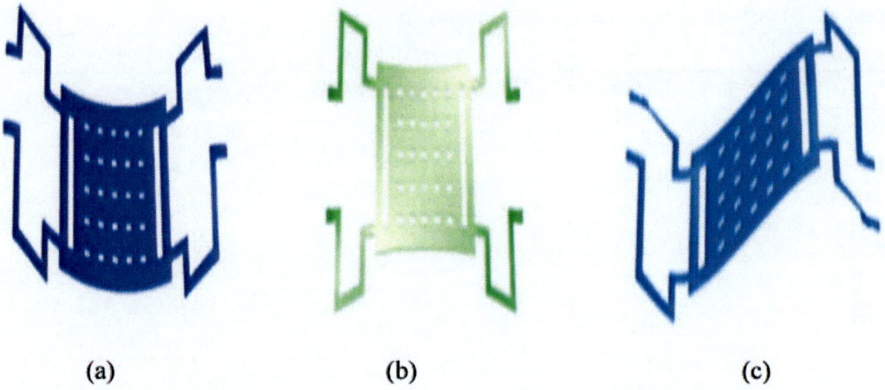

Fig. 3 Serpentine membrane resonant frequencies, **a** 4608 Hz, **b** 7772 Hz, 12244 Hz

2.2 Reconfigurable Antenna Using RF MEMS Switches

The reconfigurable microstrip patch antenna is designed by placing three identical shunt capacitive RF MEMS switches i.e., S_1, S_2, and S_3 (Fig. 4).

All three switches are individually connected to biasing pads. Silicon is used as a substrate. Silicon dioxide is used as an insulating material on the top of the silicon substrate (Figs. 5 and 6).

CPW type feeding is used with G/S/G value 50 μm/100 μm/50 μm. The structure of the reconfigurable antenna is designed by connecting the four patches i.e., P1, P2, P3, and P4. The P1 patch is conned with P2, P3, and P4 through three switches S1, S2, and S3, respectively. If S1-OFF, S2-OFF, and S3-OFF, the antenna is resonating at 5.4 GHz. If S1-OFF, S2-OFF, and S3-ON, the antenna is resonating at 0.9 GHz. If S1-OFF, S2-ON, and S3-ON, the antenna is resonating at 1.5 GHz. If S1-OFF, S2-ON, and S3-ON, the antenna is resonating at 3.5 and 8.4 GHz. Based on the switching of the RF MEMS switches, the antenna is resonating at 0.9, 1.5, 3.5, and 5.4 GHz. So, the proposed antenna is suitable for multiple applications like RFID, GPS, WiMAX, and WLAN (Table 3).

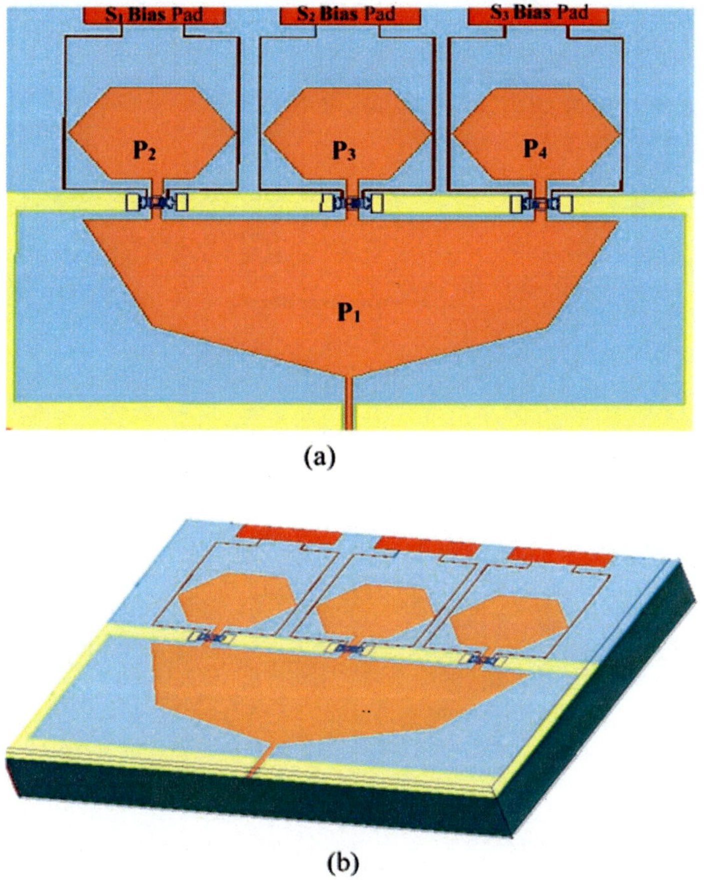

Fig. 4 Reconfigurable antenna with three identical RF MEMS switches, **a** top view, **b** side view

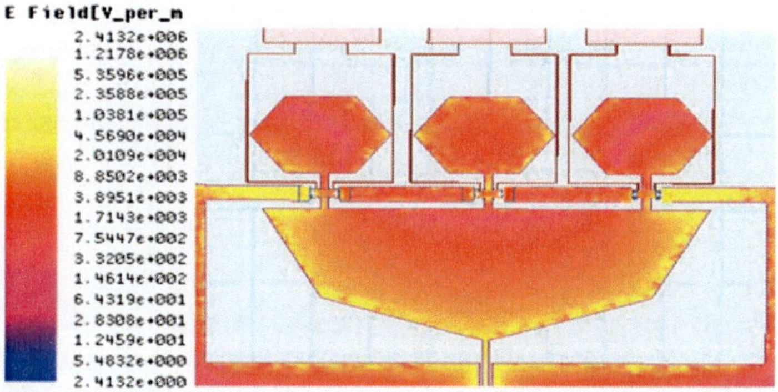

Fig. 5 Flow of current

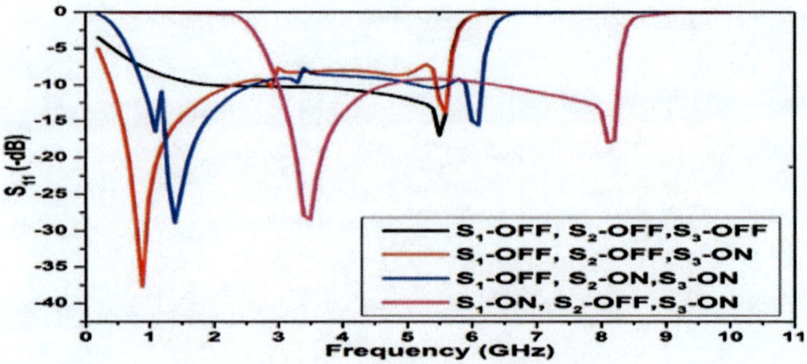

Fig. 6 Reconfigurable antenna resonating at multiple frequencies under different switching conditions

Table 3 Antenna resonating frequencies under different switching conditions

Switching condition	Antenna resonating frequency (GHz)	Application
S_1-OFF, S_2-OFF, S_3-OFF	5.4	WLAN
S_1-OFF, S_2-OFF, S_3-ON	0.9	RFID
S_1-OFF, S_2-ON, S_3-ON	1.5	GPS
S_1-OFF, S_2-ON, S_3-ON	3.5 and 8.4	WiMAX

3 Conclusion

The demand for the reconfigurable antenna is increasing day by day. In this paper, we have designed a reconfigurable microstrip patch antenna for low-frequency communication applications using shunt capacitive RF MEMS switches. The RF MEMS switch used is offering high performance. The reconfigurable antenna is designed with three RF MEMS switches. Based on the switching of the RF MEMS switches, the antenna is resonating at 0.9, 1.5, 3.5, 5.4 GHz. So, the designed antenna is suitable for multiple applications like RFID, GPS, WiMAX, and WLAN.

References

1. Ramana, R., Shanmugananthamb, T., Sindhanaiselvic, D.: Analysis of RF MEMS series switch with serpentine spring shaped cantilever beam for wireless applications. Mater. Today: Proc. **5**, 1890–1896 (2018)
2. Shekhar, S., Vinoy, K.J., Anantha Suresh, G.K.: Surface-micromachined capacitive RF switches with low actuation voltage and steady contact. J. Microelectromech. Syst. (2017). https://doi.org/10.1109/jmems.2017.2688519

3. Mulloni, V., Margesin, B., Farinelli, P., Marcelli, R., Lucibello, A., De Angelis, G.: Cycling reliability of RF–MEMS switches with Gold-Platinum multilayers as contact material. Microsyst. Technol. (2015). https://doi.org/10.1007/s00542-015-2782-2
4. Shah, I.A., Hayat, S., Basir, A., Zada, M., Shah, S.A.A., Ullah, S., Ullah, S.: Design and analysis of a hexa-band frequency reconfigurable antenna for wireless communication. Int. J. Electron. Commun. (2018). https://doi.org/10.1016/j.aeue.2018.10.012
5. Kang, H.Y., Hu, H.Y., Han, B.G., Su, H.: A reconfigurable solid-state plasma dipole antenna based on SPiN diodes. Microelectron. Eng. **214**, 55–59 (2019). https://doi.org/10.1016/j.mee.2019.04.025
6. Abou Al-Alaa, M., Elsadek, H.A., Abdallah, E.A.: Compact multi-band frequency reconfigurable planar monopole antenna for several wireless communication applications. J. Electr. Syst. Inf. Technol. **1**, 17–25 (2014). http://dx.doi.org/10.1016/j.jesit.2014.03.008
7. Cetiner, B.A., Jafarkhani, H., Qian, J.-Y., Yoo, H.J., Grau, A., De Flaviis, F.: Multifunctional reconfigurable MEMS integrated antennas for adaptive MIMO systems. Adaptive antennas and MIMO systems for wireless systems. IEEE Commun. Mag. (2004)
8. Deng, Z., Hong, M., Yao, Y.: Ka-band reconfigurable microstrip patch antenna with switchable pattern. Integr. Ferroelectr.: Int. J. **128**(1), 49–53 (2011). https://doi.org/10.1080/10584587.2011.576186
9. Gil, I., Morata, M., Fernández, R., Rottenberg, X., De Raedt, W.: Characterization and modelling of switchable stop-band filters based on RF-MEMS and complementary split ring resonators. Microelectron. Eng. **88**, 1–5 (2011)
10. Haridas, N., Erdogan, A.T., Arslan, T., Walton, A.J., Smith, S., Stevenson, T., Dunare, C., Gundlach, A., Terry, J., Argyrakis, P., Tierney, K., Ross, A., O'Hara, T.: Reconfigurable MEMS antennas. In: NASA/ESA Conference on Adaptive Hardware and Systems, pp. 147–154. IEEE Computer Society (2008). https://doi.org/10.1109/ahs.2008.28
11. Kim, I., Rahmat-Samii, Y.: RF MEMS switchable slot patch antenna integrated with bias network. IEEE Trans. Antennas Propag. **59**(12) (2011)
12. Rebeiz, G.M., Muldavin, J.B.: RF MEMS switches and switch circuits. IEEE Microw. Mag. (2001)
13. Angira, M., Bansal, D., Kumar, P., Mehta, K., Rangra, K.: A novel capacitive RF-MEMS switch for multi-frequency operation. Superlattices Microstruct. **133** (2019)
14. Sharma,· U., Kumar, M., Sharma, R., Saha, T., Jain, K.K., Dutta, S., Sharma, E.K.: Fabrication process induced changes in scattering parameters of meander type RFMEMS shunt switch. Microsyst. Technol. (2017). https://doi.org/10.1007/s00542-017-3314-z
15. Guha, K., Laskar, N.M., Gogoi, H.J., Borah, A.K., Baishnab, K.L., Baishya, S.: Novel analytical model for optimizing the pull-in voltage in a flexured MEMS switch incorporating beam perforation effect. Solid-State Electron. **137**, 85–94 (2017)
16. Thalluri, L.N., Guha, K., Srinivasa Rao, K., Venkata Hari Prasad, G., Girija Sravani, K., Sastry, K.S.R., Kanakala, A.R., Bose Babu, P.: Perforated serpentine membrane with AlN as dielectric material shunt capacitive RF MEMS switch fabrication and characterization. Microsyst. Technol. (2020). https://doi.org/10.1007/s00542-020-04755

Design, Analysis and Simulation of a Piezoresistive Microbridge and Microcantilever for MEMS Pressure Sensor in Continuous Glucose

G. Sai Lakshmi, K. Srinivasa Rao, Koushik Guha, and K. Girija Sravani

1 Introduction

They detect the blood glucose levels using the topology of interdigital electrodes with the help of COMSOL Multiphysics, this type of interdigital electrodes is used to detect the blood glucose levels by doing several tests in a non-invasive method, which is to decrease the physical injuries and infection [1]. They determine the simulation and modeling of blood pressure using a piezoelectric cantilever sensor [2]. Their research paper shows the development of a non-invasive device to monitor the blood glucose levels [3]. To detect the hypodermic continuous glucose monitoring sensor, a needle-type multi-electrode array has been fabricated using the MEMS technology [4]. They determine a sensor, a device that consists of microcantilever, which can be driven by a remote magnetic field and separated by the sensing environment using the semi-permeable membrane [5]. Their system is based on the Beer's and Lambert's principle to detect the variations in absorption of IR light by glucose levels in blood [6]. Their aim is to design a sensor, which is a non-invasive approach of biosensor used to detect the blood glucose levels using COMSOL Multiphysics [7]. They introduce a differential affinity sensor for the continuous glucose monitoring system [8]. This paper presents the different models of capacitive pressure sensors with which the capacitance can be varied with an increase in pressure [9]. They introduce a sensor that detects the continuous blood glucose levels with the help of micro-cantilever and a semi-permeable membrane [10]. The two devices microcantilever and membrane situated inside the microchambers are used to detect glucose levels in the blood [11].

G. Sai Lakshmi · K. Srinivasa Rao (✉) · K. Girija Sravani
Department of Electronics and Communication Engineering, MEMS Research Center, KL University, Green Field, Vaddeswaram, Guntur 522502, India
e-mail: drksrao@kluniversity.in

K. Guha · K. Girija Sravani
Department of ECE, National MEMS Design Centre, National Institute of Technology, Silchar, Silchar, Assam, India

A. Biswas et al. (eds.), *Microelectronics, Circuits and Systems*, Lecture Notes in Electrical Engineering 755, https://doi.org/10.1007/978-981-16-1570-2_22

Blood glucose sensor is injected into the person, if they inject insulin into the human body it shows the readings through LCD and mobiles screen [12]. The microwave spectroscopy is attached to the interdigital capacitive sensor to detect blood glucose levels through electromagnetic properties [13]. They present a MEMS-based affinity-sensor via dielectric properties [14]. They proposed a capacitive pressure sensor using MEMS technology for continuous glucose monitoring in diabetes treatment [15]. References [16–19] shows the different structures of pressure sensors with different types of mechanisms. They proposed a multi-channel nanocavity-coupled biosensor with waveguides and nanocavity in the hexagonal lattice structure in the air slab to detect the concentration of glucose in blood [20]. They focus on the different types of capacitive pressure sensor principles, which simulate the capacitance when pressure is applied [21]. They used the electro-osmotic pressure sensor, which is used to measure the glucose levels in the blood [22]. The application for drug delivery system, micropump, is designed and evaluated [23–25]. The design of microbridges is done using three principles piezoresistive, piezoelectric and capacitive [26]. To determine the performance of the capacitive pressure sensor, the two plates are separated by using a dielectric medium [27]. The capacitive pressure sensor is designed for the harsh environment [28]. The MEMS-based piezoelectric micropump is designed and its performance can be calculated by using the FEM tool [29]. To calculate the reliability of the device, stress analysis is performed on the piezoelectric micropump [30]. The capacitive-based microstructure is designed to detect Parkinson's disease [31]. Human kidneys have a tendency to get affected easily, so the artificial kidney is needed [32–34]. The design, analysis, simulation, optimization of the MEMS piezoelectric micropump is used in the application of drug delivery system [35].

As of now, two types of approaches are presented, one is non-invasive and the other is minimally-invasive. Non-invasive technique observation is seen by optical devices by pointing a light ray into skin to determine properties in reflected light, minimally-invasive measures the change in glucose levels that is either consumption of oxygen or production of hydrogen peroxide. Here, we proposed a sensor with high sensitivity and better time response.

The rest of the sections are as follows, the proposed device, the working principle and theoretical parameters are discussed in Sect. 2, optimized analysis and the parametric analysis are discussed in Sect. 3 and the conclusions are discussed in Sect. 4.

2 Methodology

2.1 Proposed Device

The proposed device has three piezoresistors that are placed on top of the membrane and using p-type single crystalline silicon. The pressure sensor has a square cavity loaded with a standard solution of reference concentration, a semi-permeable

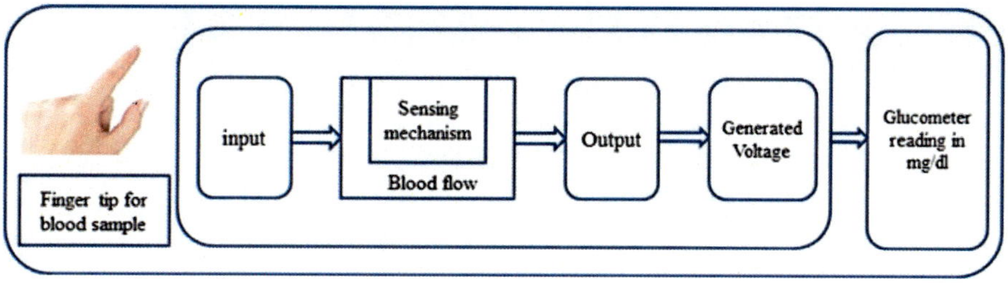

Fig. 1 Proposed device

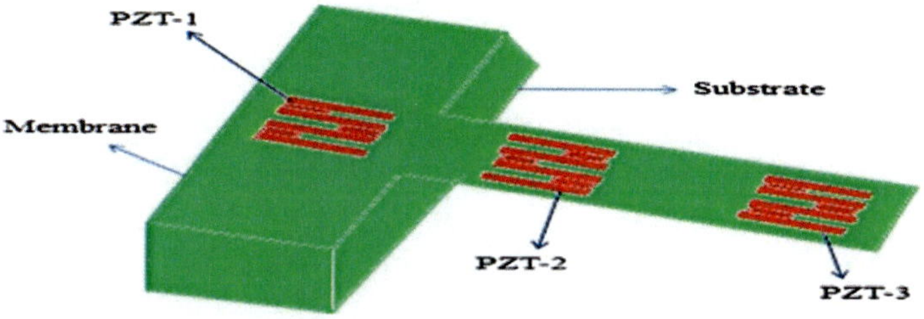

Fig. 2 Schematic view of microcantilever

membrane is utilized to close the filled cavity and silicon membrane is kept on the top of the cavity. Inside the square cavity, 100 mg/dl is placed and in the test chamber, the concentration varies from 50 to 450 mg/dl. When the fluid is detected outside, osmotic pressure creates across the semi-permeable membrane, and then it gives the deformation in microcantilever in terms of change in resistance in the form of the output voltage (Figs. 1 and 2).

In this paper, microcantilever is constructed, one end of the beam is fixed and the other end of the beam is movable. It can be designed using the piezoresistors, one is at the substrate and another is at the length of the microcantilever and is placed at the edges of the microcantilever as maximum stress is at the edges compared to the middle. Three membrane materials, silicon dioxide, aluminum nitride, silicon nitride, and three piezoresistive materials, P-type single crystalline silicon, N-type single crystalline silicon and PZT-5H, are used in the microcantilever. The deformation in the microcantilever is seen by applying load across the length of a microcantilever and measures in the form of change in resistance and the form of the output voltage.

2.2 *Working Principle*

In this paper, the glucose sensor uses the osmosis principle to detect the change in glucose concentration levels. Osmosis is defined as the solvent, flow across the semi-permeable membrane driven by a difference in concentration levels, then the osmotic pressure develops and molecules move from lower concentration to higher concentration.

2.3 Theoretical Parameters

Resistance can be calculated as

$$\mathrm{R} = \rho_0 \frac{L_R}{W_R T_R} \tag{1}$$

where ρ_0 is the resistivity of the material, L_R, W_R, T_R are the length, the width and the thickness of the piezoresistor.

Stress is given by

$$\sigma_l = \frac{M.y}{l} \tag{2}$$

σ represents the stress, suffix l refers to longitudinal stress, y is the distance between the natural axis to bending axis M is bending moment.

Change in resistance is given as

$$\frac{\Delta R}{R} = \pi_l \sigma_l. \tag{3}$$

Here $\Delta R/R$ represents the change in resistance, π is the piezoresistive coefficient, σ is the applied stress and suffix l refers to the longitudinal stress for resistor axis.

3 Results and Discussions

3.1 Optimized Analysis

The length of the microcantilever increases then non-linearity also increases, it is smaller for lower lengths. Figure 3 shows the non-linearity of the microcantilever when the force is applied from 0 to 2.5 μN at different lengths from 300 to 500 μm for silicon dioxide membrane. Figure 4 shows the non-linearity of the microcantilever when the force is applied from 0 to 2.5μN at different lengths from 300 to 500 μm for aluminum nitride membrane. Figure 5 shows the non-linearity of the microcantilever when the force is applied from 0 to 2.5 μN at different lengths from 300 to 500 μm for silicon nitride membrane. The optimized length and width can be taken from non-linearity and sensitivity aspects.

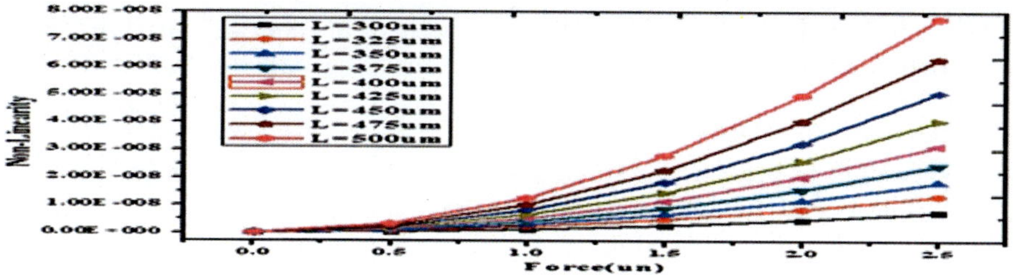

Fig. 3 Force versus non-linearity for SiO_2 membrane at different lengths

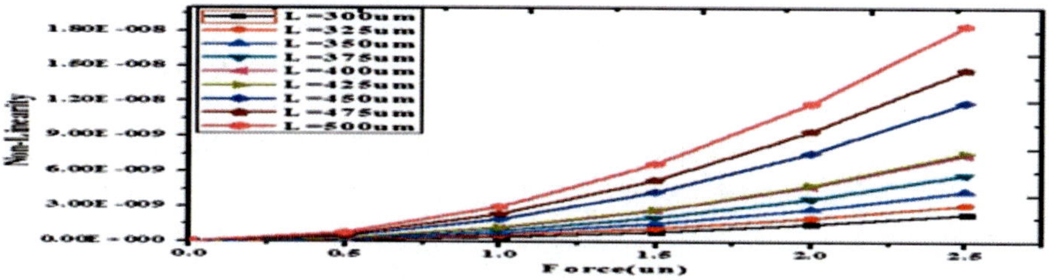

Fig. 4 Force versus non-linearity for ALN membrane at different lengths

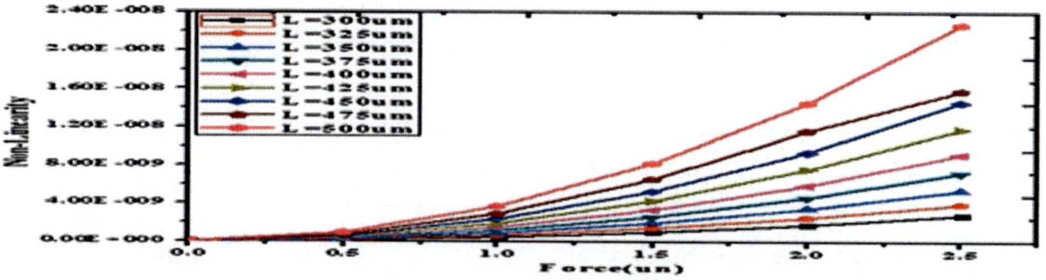

Fig. 5 Force versus non-linearity for Si_3N_4 membrane at different lengths

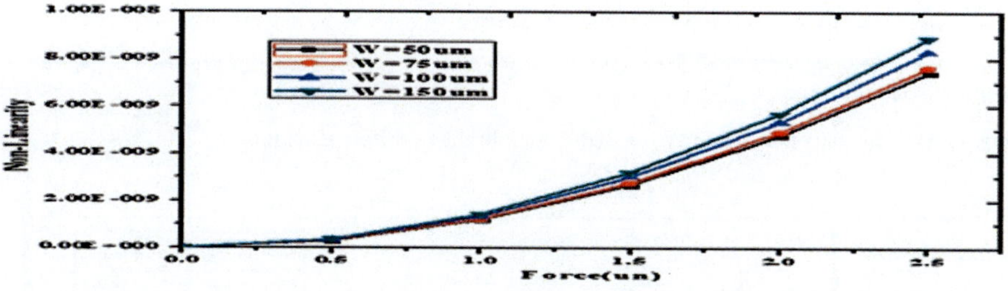

Fig. 6 Force versus non-linearity for SiO_2 membrane at different widths

The width of the microcantilever increases then non-linearity also increases, Fig. 6 shows a non-linearity of a microcantilever when the force is applied from 0 to 2.5μN at different widths from 50 to 150 μm for silicon dioxide membrane. Figure 7 shows a non-linearity of the microcantilever when the force is applied from 0 to 2.5μN at different widths from 50 to 150 μm for aluminum nitride membrane. Figure 8 shows the non-linearity of the microcantilever when the force is applied from 0 to 2.5μN at different widths from 50 to 150 μm for silicon nitride membrane.

Piezoresistive thickness has more impact in sensitivity rather than piezoresistive width, Fig. 9 shows the sensitivity of the microcantilever when the Pzt thickness is varying from 0.2 to 2 μm and Pzt width is varying from 4 to 2 μm for P-type single crystalline silicon material. Figure 10 shows the sensitivity of the microcantilever when the Pzt thickness is varying from 0.2 to 2 μm and Pzt width is varying from 4 to

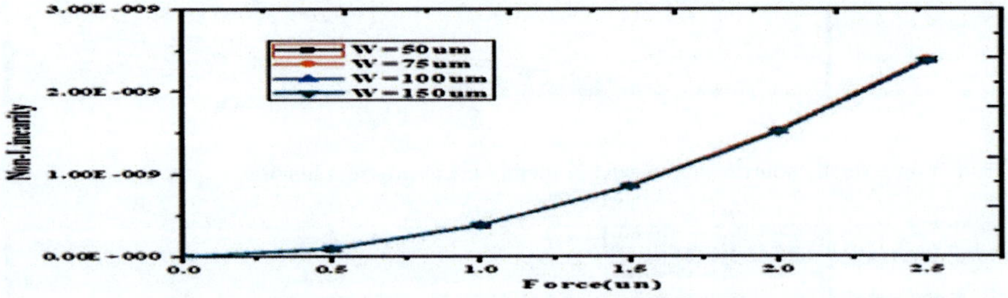

Fig. 7 Force versus non-linearity for ALN membrane at different widths

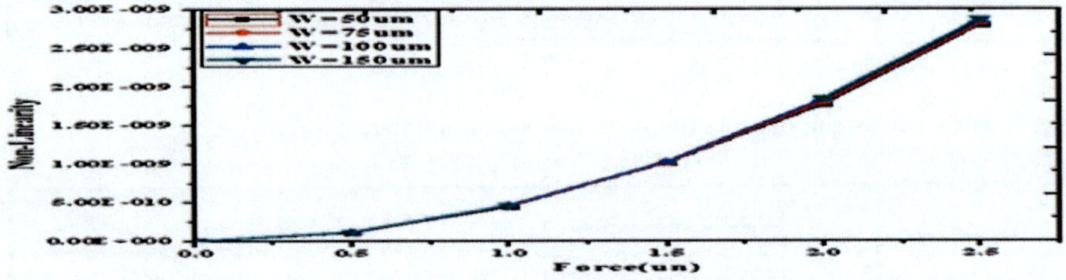

Fig. 8 Force versus non-linearity for Si_3N_4 membrane at different widths

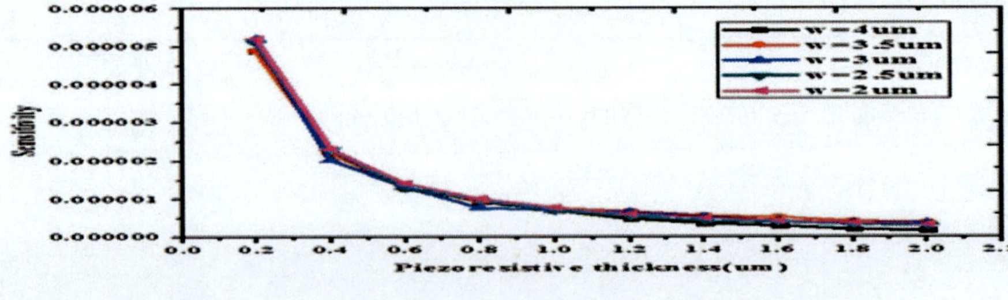

Fig. 9 Piezoresistive thickness versus sensitivity at various widths for P-type silicon

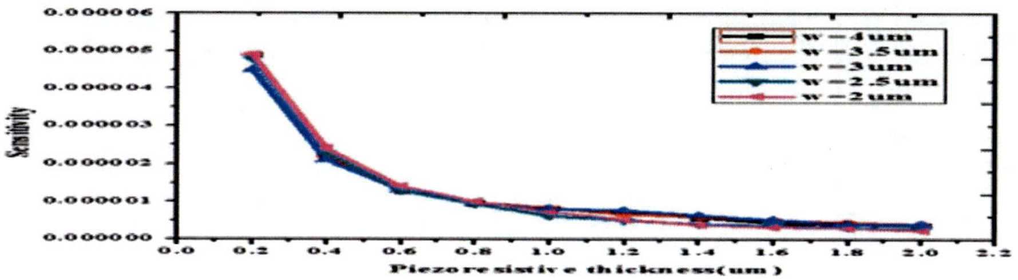

Fig. 10 Piezoresistive thickness versus sensitivity at various widths for N-type silicon

2 μm for N-type single crystalline silicon material. P-type single crystalline silicon has more sensitivity compared to the N-type single crystalline silicon.

The unique characteristics of a microcantilever are that it bends when the molecular adsorption is confined to one side of the surface. The deflection in the microcantilever is obtained by applying uniformly distributed load along the length of the microcantilever, which measures the surface stress by the adsorption of molecules and also measures the change in resistance as shown in Fig. 11. The deflection in the microcantilever when applying the force ranging from 0 to 2.5μN is shown in Fig. 12.

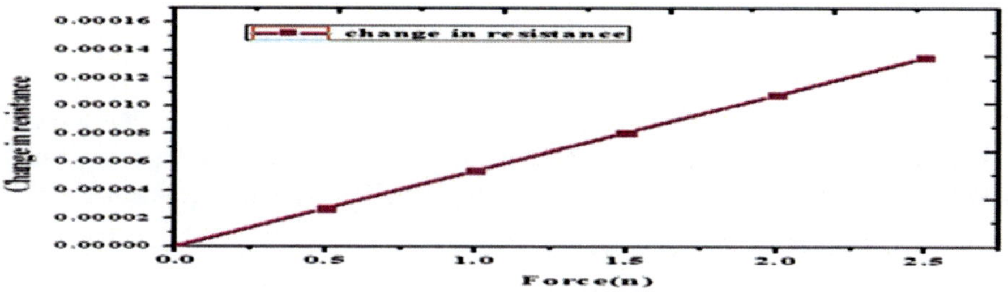

Fig. 11 Force versus change in resistance

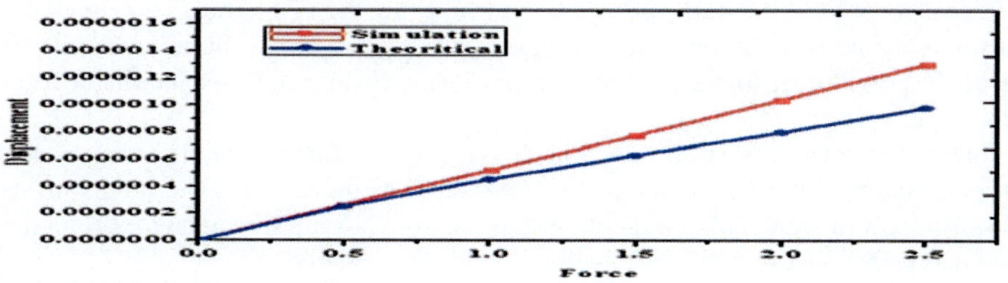

Fig. 12 Force versus displacement

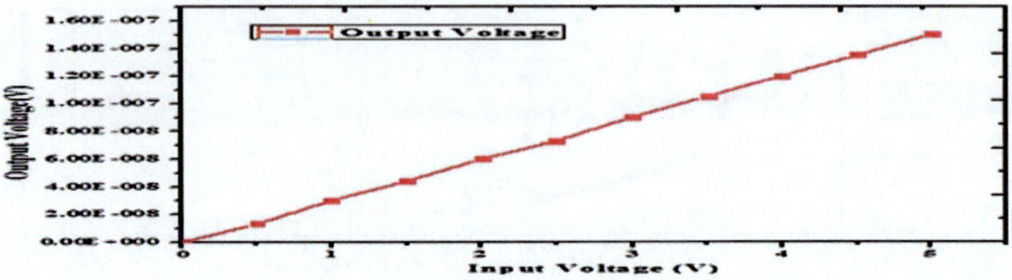

Fig. 13 Relationship between the input voltage and output voltage

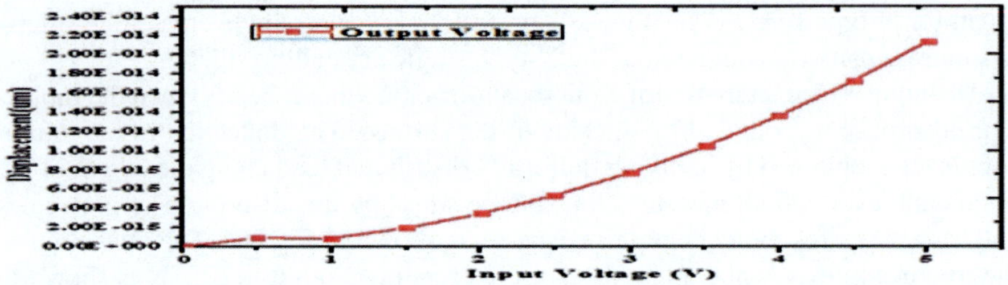

Fig. 14 Input voltage versus displacement

3.2 Parametric Results

A minimum difference in change in resistance may vary the stress in the form of voltage. Figure 13 shows a relationship between the input voltage and output voltage and Fig. 14 shows the relation between the input voltage and displacement.

4 Conclusion

The microcantilever is designed and optimized using the FEM Tool. Three piezoresistors are placed on the top of the microcantilever, by integrating this microcantilever with the pressure sensor then we could calculate the Change in Resistance in the form of output voltage.

Different materials SiO_2, AlN and Si_3N_4 are used for membrane in microcantilever, among them silicon dioxide is considered as membrane material due to its low young's modulus. Materials like P-type silicon, N-type silicon and PZT-5H are used as piezoresistive materials. P-type silicon is considered as Pzt material because P-type has more sensitivity than the N-type. Silicon, glass (Quartz), LCP are used as substrate materials, owing to its high-cost silicon is taken as substrate material. The optimized dimensions 400 μm, 50 μm, 1 μm are length, width and thickness of membrane. Piezoresistive width and piezoresistive thickness are 2 and 0.2 μm.

Table 1 Comparison between the microbridge and microcantilever

Component	Microbridge	Microcantilever
Sensitivity	1.5e−7	5.2e−6
Change in resistance	0.000015	0.00011
Output voltage	5.11 μv	0.15 μv
Displacement	2.2e−4 m	1.04 μm

Microcantilever having more parametric values than the microbridge as shown in Table 1. So, the microcantilever is further integrated into the pressure sensor.

Acknowledgements The authors would like to thank NMDC supported by NPMASS, National Institute of Technology, Silchar for providing the necessary computational tools.

References

1. Pockevicius, V., Markevicius, M., Cepenas, D., Andriukaitis, D.: Navikas: blood glucose level estimation using interdigital electrodes. Elektronika Ir Elektrotechnika **19**(6) (2013). ISSN 1392-1215
2. Panwar, L.S., Panwar, V., Kala, S., Panwar, S.S., Sharma, S.: Design of MEMS piezoelectric blood pressure sensor. 978-15090-6403-8/17. IEEE (2017)
3. Nandi, B., Mondal, P., Chowdhury, S.R.: A Non-invasive blood insulin and glucose monitoring system based on Near-Infrared spectroscopy with remote data logging. In: IEEE 31st International Symposium on Computer-Based Medical Systems, 2372-9198/18. IEEE (2018)
4. Jung, M.W., Kim, D.W., Jeong, R.A., Kim, H.C.: Needle-type multi-electrode array fabricated by MEMS technology for the hypodermic continuous glucose monitoring system. In: 26th Annual International Conference of the IEEE EMBS, 0-7803-8439-3/04. IEEE (2004)
5. Huang, X., Li, S., Schultz, J., Wang, Q., Lin, Q.: A Mems sensor for continuous monitoring of glucose in subcutaneous tissue. 978-1-4244-2978-3. IEEE (2009)
6. Faris, M., Ansari, T.A., Javid, T., Haseeb, A., Mujib, M.D.: Towards design and development of noninvasive glucose monitoring system. 978-1-5090-3310-2/17. IEEE (2017)
7. Bhawiskar, S., Manza, R.R., Dhole, B.N.: Simulating an IDC-bio-sensor to detect diabetes. Int. J. Sci. Res. **3**(12) (2014)
8. Huang, X., Oxsher, J., LeDuc, C., Ravussin, Y., Wang, Q., Accili, D., Leibel, R., Lin, Q.: A MEMS differential affinity sensor for continuous glucose detection. 978-1-4577-0156-6/11. IEEE (2011)
9. Ameta, S., Sharma, A., Inaniya, P.K.: Designing a multichannel nanocavity coupled photonic crystal biosensor for detection of glucose concentration in blood. In: 8th ICCCNT 2017, IEEE July 3–5. IIT Delhi, India (2017)
10. Zhao, Y.J., Davidson, A., Bain, J., Li, S.Q., Wang, Q., Lin, Q.: A MEMS viscometric glucose monitoring device. In: The 13th International Conference on Solid-State Sensors. IEEE (2005)
11. Huang, X., Li, S., Schultz, J., Wang, Q., Lin, Q.: A biocompatible affinity MEMS sensor for continuous monitoring of glucose. In: 4th IEEE International Conference on Nano/Micro Engineered and Molecular Systems (2009)
12. Jeyapriya, S., Ramalakshmi, R.: Glucose monitoring and control of diabetes using GSM and automatic insulin injector system for human bodies. In: IEEE International Conference on Intelligent Techniques in Control, 978-1-5090-4778-9/17. IEEE (2017)

13. Morshidi, W.H.W., Zaharudin, Z., Khan, S., Nordin, A.N., Shaikh, F.A., Adam, I., Kader, K.A.: Inter-digital sensor for non-invasive blood glucose monitoring. In: IEEE International Conference on Innovative Research and Development (ICIRD), 978-1-5386-5283-1/18. IEEE (2018)
14. Huang, X., LeDuc, C., Ravussin, Y., Li, S., Song, B., Wang, Q., Accili, D., Leibel, R., Lin, Q.: A MEMS differential dielectric sensor for continuous glucose monitoring. 978-1-4673-0325-5/12. IEEE (2012)
15. Samyuktha, N., Maneesha, P., Sreelakshmi, B.R., Pattnaik, P.K., Narayan, K.: Application of MEMS based capacitive sensor for continuous monitoring of glucose. 978-1-4799-8641-5/15. IEEE (2015)
16. Li, C., Cordovilla, F., Jagdheesh, R., Ocaña, J.L.: Design and optimization of a novel structural MEMS piezoresistive pressure sensor. Microsyst. Technol. (Springer) (2016). https://doi.org/10.1007/s00542-016-3187-6
17. Santosh Kumar, S., Pant, B.D.: Design of piezoresistive MEMS absolute pressure sensor. In: 16th International Workshop on Physics of Semiconductor Devices, vol. 8549, 85491G (2013)
18. Al Mahruz, T., Matin, R., Wahid, F.B., Dev, T.: Material and performance analysis of MEMS piezoresistive pressure sensor. Int. J. Eng. Trends Technol. (IJETT) **31**(1) (2016)
19. Santosh Kumar, S., Pant, B.D.: Design principles and considerations for the 'ideal' silicon piezoresistive pressure sensor. Microsyst. Technol. (2014)
20. Ameta, S., Sharma, A., Inaniya, P.K.: Designing a multichannel nano cavity coupled photonic crystal biosensor for detection of glucose concentration in blood. In: 8th ICCCNT July 3–5, IIT Delhi. IEEE (2017)
21. Kumar, K., Balavalad, B., Sheeparamatti, B.G.: MEMS capacitive pressure sensors. Sens. Transducers **187**(4) (2015)
22. Girija Sravani, K., Sai Lakshmi, G., Srinivasa Rao, K.: Design and optimization of piezoresistive materials based microbridge for electro-osmosis pressure sensor. Trans. Electr. Electron. Mater. **20**, 198–205 (2019)
23. Srinivasa Rao, K., Sateesh, J., Girija Sravani, K., Ashok Kumar, P., et al.: Design and analysis of MEMS based piezoelectric micro pump integrated with micro needle. Trans. Electr. Electron. Mater. **26**, 3153–3159 (2018)
24. Srinivasa Rao, K., Sateesh, J., Girija Sravani, K., Guha, K.: Design and flow analysis of MEMS based piezo-electric micro pump. Microsyst. Technol. 1609–1614 (2017)
25. Sateesh, J., Srinivasa Rao, K.: Simulation of MEMS based drug delivery system. J. Biosens. Bioelectron. Mater. Sci. (2017)
26. Girija Sravani, K., Sai Lakshmi, G., Srinivasa Rao, K.: Design of piezoresistive based micro bridge for MEMS pressure sensor. Micro Nanosyst. **12**(May) (2019)
27. Srinivasa Rao, K., Girija Sravani, K., Ashok Kumar, P., Samyuktha, W., Vazad Vardhan, D., Girish Naidu, B., Guha, K.: Design and sensitivity analysis of capacitive MEMS pressure sensor for blood pressure measurement. Microsyst. Technol. (2020)
28. Srinivasa Rao, K., Girija Sravani, K., Ashok Kumar, P., et al.: Design and simulation of MEMS based capacitive pressure sensor for harsh environment. Microsyst. Technol. (2020)
29. Srinivasa Rao, K., Girija Sravani, K., Ashok Kumar, P., Hamza, Md., et al.: Design and performance analysis of PDMS based micropump. Trans. Electr. Electron. Mater. (2020)
30. Srinivasa Rao, K., Girija Sravani, K., Ashok Kumar, P., Hamza, Md., et al.: Design and optimization of MEMS based piezoelectric actuator for drug delivery systems. Microsyst. Technol. (2019)
31. Srinivasa Rao, K., Prasad, G.R.K., Siddaiah, N., Srinivas Babu, P.S.: Design and simulation MEMS based sensor for early detection of PD. In: International Conference on Electrical, Electronics, Communication, Computer and Optimization Techniques (ICEECCOT). IEEE (2017)
32. Sateesh, J., Srinivasa Rao, K., Guha, K., et al.: Recreating the size dependent re-absorption function of proximal convoluted tubule towards artificial kidney applications—structural analysis and computational study. Artif. Organs **44**(113) (2020)

33. Sateesh, J., Srinivasa Rao, K., Guha, K., et al.: Regenerating re-absorption function of proximal convoluted tubule using microfluidics for kidney-on-chip applications. SN Appl. Sci. **2**(1) (2020)
34. Guha, K., Sateesh, J., Srinivasa Rao, K., et al.: Mimicking kidney re-absorption using microfluidics by considering hydrostatic pressure inside kidney tubules: structural and analytical study. Microsyst. Technol. **26**(10) (2019)
35. Guha, K., Sateesh, J., Srinivasa Rao, K., et al.: Design and optimization of MEMS based piezo electric-micropump. In: 4th International Conference on Microelectronics, Circuits and Systems (MICRO), Darjeeling, West Bengal, India (2017)

Investigation of the Optoelectronic Properties of InSbNBi with 16-Band k Dot p Model

Indranil Mal and Dip Prakash Samajdar

1 Introduction

The advent of sophisticated and state-of-the-art semiconductor alloy fabrication techniques such as molecular beam epitaxy, metal organic chemical vapor deposition and pulsed laser ablation have offered the opportunity to explore the dilute impurity containing ternary and quaternary semiconductors with promising properties [1, 2]. Dilute Bismuth and Nitrogen-containing III–V narrow band gap semiconducting materials grown over the commonly used III–V substrates drew the attention of infrared community from the last few years due to their fascinating optoelectronic properties in the mid and far-infrared regime [3]. InSb, being a widely used binary semiconductor for mid-wave infrared (MWIR) regime [4], it is incapable to operate in the long wavelength infrared (LWIR) regime. Dilute incorporation of impurities like Bi and N can address the issue [4]. Solely N or Bi impurity perturbs the conduction and valence band of the host InSb respectively, which drives the host further towards the LWIR regime. However, the resultant ternary material (InSbN/InSbBi) undergoes several issues like lots of defects in the material due to lattice mismatch with the host, loss of stoichiometry of the host and inefficient optoelectronic devices. Co-incorporation of Bi and N helps to achieve the lattice-matched condition with the host material, as a result, it can mitigate these issues too. In order to achieve a specific operating wavelength in LWIR window, in case of co-incorporation of N and Bi the total amount of impurity is very less as compared with the single Bi or N incorporation. Here, we have modeled the 16-band k·p perturbation Hamiltonian [5, 6], considering the Conduction Band Anticrossing (CBAC) [2, 7] and Valence Band Anticrossing (VBAC) [8] model to taking into account the effect of N and Bi incorporation, respectively. Finally, we have calculated the optoelectronic properties of $InSb_{1-x-y}N_xBi_y$.

I. Mal (✉) · D. P. Samajdar
Department of Electronics & Communication Engineering, PDPM Indian Institute of Information Technology Design and Manufacturing, Jabalpur, India

A. Biswas et al. (eds.), *Microelectronics, Circuits and Systems*, Lecture Notes in Electrical Engineering 755, https://doi.org/10.1007/978-981-16-1570-2_23

2 Mathematical Modeling

N impurity creates isoelectronic defect state 0.47 eV above the conduction band edge of host atom, whereas the Bi defect states lie 1 eV below the valence band maxima of InSb. These isoelectronic defect states interact and split up the band edges of host atom into bonding and antibonding sub-bands, respectively. Consequently, a lowering in band gap is accomplished. The smaller size N impurity creates tensile strain and larger size Bi creates compressive strain, as a result, the strain compensation can be achieved by co-incorporation. The ratio is 0.14 between the N and Bi concentration for which strain compensation is achieved. The host Hamiltonian ($H_{8\times8}$) is modeled accordingly [5, 9], where the diagonal terms correspond to the respective sub-bands extrema, and all other terms are elaborated in Ref. [7]. Finally, including the impurity and coupling matrix components with the system Hamiltonian, the $H_{16\times16}$ has been formulated as follows:

$$H_{16\times16} = \begin{bmatrix} H_{8\times8} & V_{N,Bi(8\times8)} \\ V_{N,Bi(8\times8)} & H_{N,Bi(8\times8)} \end{bmatrix} \tag{1}$$

The readers are encouraged to go through [7, 8] for the formulation of the 16-band Hamiltonian in more detail. Solving the 16-band system Hamiltonian, we have derived the reduced band gap equation for the $InSb_{1\text{-}x\text{-}y}N_xBi_y$ as a function of impurity concentrations, which is controlled by the host conduction and valence band extrema $\left(E_{CB/VB}^{InSb}\right)$, impurity energy levels $\left(E_{N/Bi}\right)$, band-offsets $(\Delta E_{VB}(x, y))$ between the binary elements as a function of impurity concentrations (x, y) and the coupling coefficients $\left(C_{N/Bi}\right)$ are as follows

$$E_g^{InSbNBi}(x, y) = \begin{bmatrix} \frac{1}{2}\left[E_{CB}^{InSb} + \Delta E_{CB}(x, y) + E_N - \sqrt{(E_{CB}^{InSb} - E_N)^2 + 4C_N^2 x}\right] \\ -\frac{1}{2}\left[E_{VB}^{InSb} + \Delta E_{VB}(x, y) + E_{Bi} - \sqrt{(E_{VB}^{InSb} - E_{Bi})^2 + 4C_{Bi}^2 y}\right] \end{bmatrix} \tag{2}$$

Computing the band gap, band offsets and effective mass, we have calculated the optical gain for $InSb_{1\text{-}x\text{-}y}N_xBi_y/InSb$ quantum well (QW) system accordingly [10] as a function of impurity concentration, surface carrier density and thickness of the QW. All the required binary material parameters are considered from Ref. [11].

3 Results and Discussion

Formulating the 16 × 16-system Hamiltonian at room temperature and one atmospheric pressure, we have calculated the eight doubly spin-degenerate Eigen values of the corresponding boning and antibonding sub-bands of $InSb_{0.9772}N_{0.0028}Bi_{0.02}$ along the highly symmetric directions with respect to the Brillouin zone center (Γ) point.

3.1 Electronic Properties

The impurity incorporation into the host material is very less: ~2% Bi and around 0.3% N, which helps to achieve a narrow band gap of 68 meV and corresponding operating wavelength of around ~18.23 μm is achieved without disturbing the stoichiometry of the host semiconductor. The co-incorporation of Bi and N helps to realize two important conditions: first, the derived material maintains strain relaxation condition with respect to InSb substrate and second, very less amount of impurity is required to achieve LWIR regime operating wavelength, which resolves one of the prior concerns of the Bi and N incorporation in III–V semiconductors. The energy dispersion along the crystallographic symmetric directions **(a)** <100> **(b)** <110> **(c)** <111> with respect to the Brillouin zone center (Γ) point are shown in Fig. 1.

It can be observed from the figure that the heavy hole (HH) and light hole (LH) sub-bands preserve degeneracy at (Γ) point, which is an evidence of strain relaxation of the derived material with respect to the host InSb. It can also be concluded from the dispersion relations that the direct band gap is very low as compared with the spin–orbit coupling splitting energy gap, which helps to mitigate the non-radiative recombination processes such as Auger recombination in optoelectronic devices. In order to diminish several fabrication challenges caused by the bigger size Bi and smaller size N as compared with host atoms, we have thoroughly studied the effect of impurity concentration. Therefore, any suitable concentration for growth process can be considered to obtain the desired band gap of the material. Furthermore, the application window corresponding to any particular band gap has also been studied and these are collectively shown in Fig. 2. It represents a near about linear variation of band gap and corresponding operating wavelength as a function of Bi and N.

A wide range of band gap starting from 170 to 40 meV and the corresponding wavelength of 7 to 30 μm can be attained, which makes InSbNBi a potential candidate

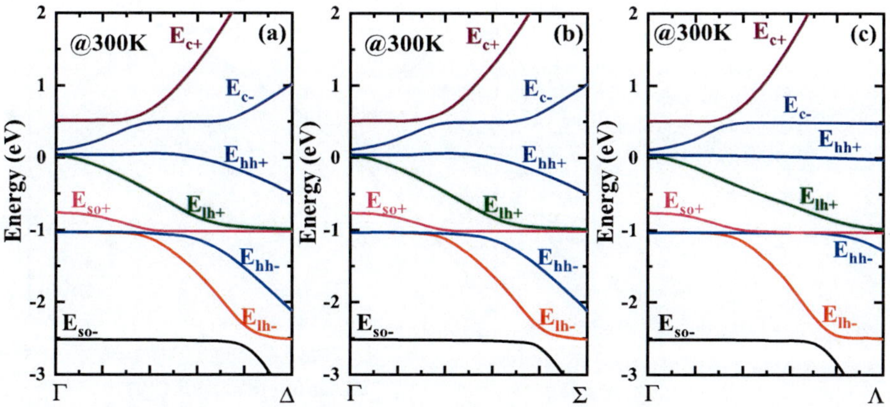

Fig. 1 Electronic band diagram of $InSb_{0.9772}N_{0.0028}Bi_{0.02}$ along three highly symmetric directions **a** 100, **b** 110, **c** 111 with respect to the Γ point at room temperature

Fig. 2 Variation of direct band gap and corresponding operating wavelength of $InSb_{1-x-y}N_xBi_y$ as a function of Bi and N concentrations

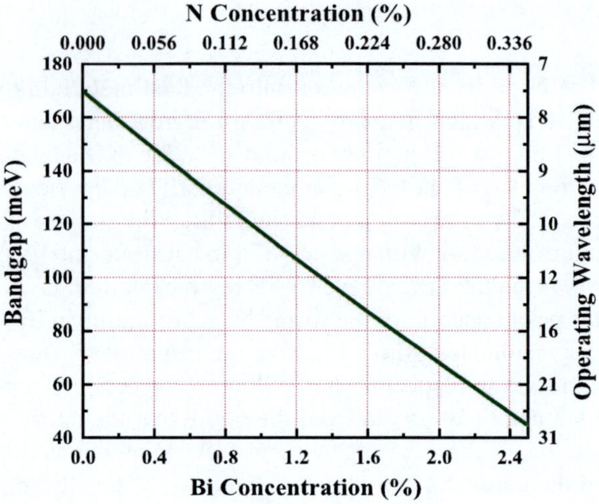

for thermal detection in LWIR region. In order to study the heterostructure, the knowledge of band offset is pivotal. The conduction and valence band offset and offset ratios of $InSb_{0.9772}N_{0.0028}Bi_{0.02}$ with respect to InSb are (ΔE_{CB}) 67.4 meV, (ΔE_{VB}) 34.5 meV, (Q_C) 66.1% and (Q_V) 33.9%, respectively. The offsets and their ratios can be tuned precisely as a function of N and Bi concentrations, which is shown in Fig. 3.

Fig. 3 Variation of conduction (in red line) and valence (in olive line) band offset ratio as a function of Bi and N concentrations

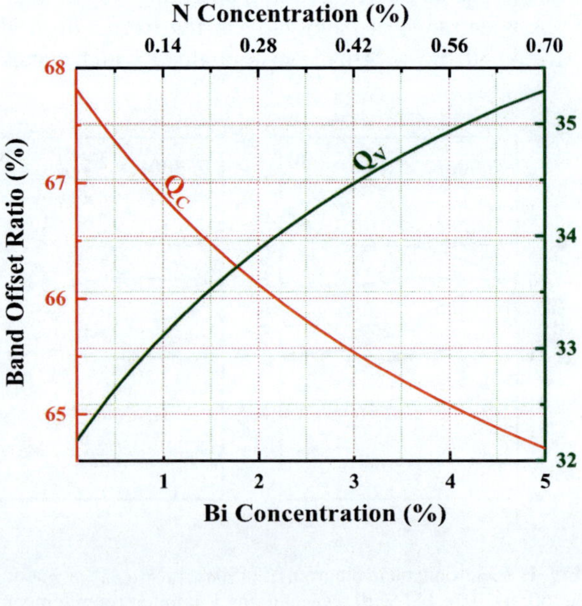

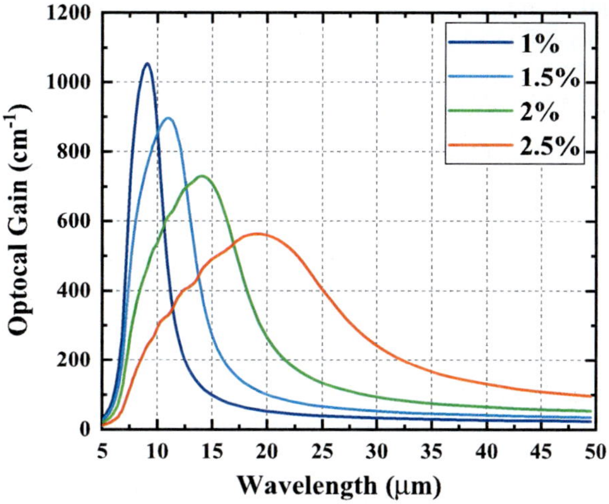

Fig. 4 Parametric variation of optical gain of InSbNBi/InSb heterostructure as a function of incident photon wavelength and Bi concentration

3.2 Optical Properties

With the help of the calculated electronic properties like band gap, band offset, effective mass, we have thoroughly investigated the optical gain for the proposed InSbNBi/InSb heterostructure, which gives a complete and precise optical gain spectra suitable for several wavelength application windows. Increasing the concentration of impurities results in reduction in band gap, which is reflected in terms of redshift in the optical spectra of the optoelectronic devices.

It can be concluded from Fig. 4 that on increasing the N and Bi concentration, optical gain spectra undergo a redshift. Along with the redshift, the FWHM of the gain spectra increase with increasing impurity concentration. Increment of FWHM is not favorable for monochromatic optoelectronic device applications.

An operating wavelength window of 9–20 μm and maximum gain value from 1000 to 500 cm^{-1} can be achieved for InSbNBi/InSb QW with a Bi concentration of 1–2.5%, 50 nm active layer thickness and surface carrier density of 4.5×10^{15} cm^{-2}. Increasing surface carrier density helps to enhance the number of populated carriers in the higher energy band, which results in an increase in the number of radiative transitions and carrier lifetime. Consequently, the maximum gain and FWHM are augmented in the gain spectra, as shown in Fig. 5. The maximum gain of $InSb_{0.9772}N_{0.0028}Bi_{0.02}$/InSb QW system varies from 450 to 950 cm^{-1} for a surface carrier density of 3×10^{15} cm^{-2} to 6×10^{15} cm^{-2}. The optical gain spectra can also be tuned by changing the active layer thickness of the QW structure. Thicker active layer leads to poor carrier confinement and smaller transition energies. As a result, on enhancing the QW thickness, the optical gain spectra undergo a redshift and smaller maximum gain value, which is evident from Fig. 6. Owing to poor confinement, the FWHM will also increase for thicker active-layered heterostructure.

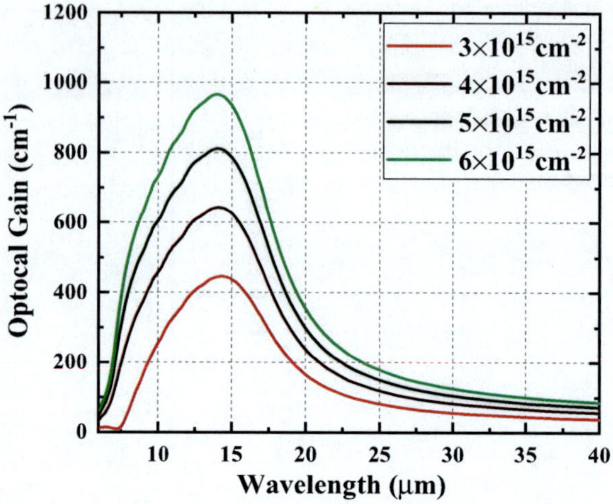

Fig. 5 Variation of optical gain of InSbNBi/InSb heterostructure as a function of incident photon wavelength and surface carrier density

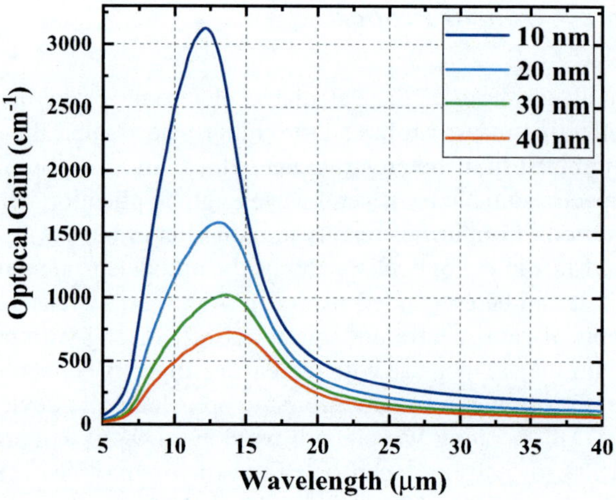

Fig. 6 Parametric variation of optical gain of InSbNBi/InSb QW as a function of incident photon wavelength and thickness of the QW layer

Therefore, depending on the requirement of the application window in LWIR regime, our proposed InSbNBi/InSb material system will be a potential candidate for optoelectronic devices.

4 Conclusions

To conclude, a suitable and promising candidate for LWIR regime in the form of InSbNBi is proposed with superior structural, electronic and optical properties at room temperature. A wide range of band gap tunability can be accomplished with a precise variation of low impurity concentration of Bi and N, which furthermore helps to realize a wide range of operating wavelength in deep IR regime. Dilute impurity concentration also helps to retain the stoichiometry of the host material. Eventually, the proposed material, InSbNBi, could be a prospective candidate for thermal image sensors with greater efficiency.

Acknowledgements Author Indranil Mal is the recipient of MHRD fellowship for carrying out his PhD work at IIITDM, Jabalpur. The authors would also like to acknowledge the Faculty Initiation Grant received from PDPM IIITDM, Jabalpur for their research on III–V material-based photodetectors.

References

1. Ng, S.T., Fan, W.J., Dang, Y.X., Yoon, S.F.: Comparison of electronic band structure and optical transparency conditions of InxGa1-xAs1-yNy/GaAs quantum wells calculated by 10-band, 8-band, and 6-band k·p models. Phys. Rev. B Condens. Matter Mater. Phys. **72**, 1–13 (2005). https://doi.org/10.1103/PhysRevB.72.115341
2. Alberi, K., Wu, J., Walukiewicz, W., Yu, K.M., Dubon, O.D., Watkins, S.P., Wang, C.X., Liu, X., Cho, Y.J., Furdyna, J.: Valence-band anticrossing in mismatched III-V semiconductor alloys. Phys. Rev. B Condens. Matter Mater. Phys. **75**, 1–6 (2007). https://doi.org/10.1103/PhysRevB.75.045203
3. Samajdar, D.P., Dhar, S.: Estimation of Bi induced changes in the direct E0 band gap of III–V-Bi alloys and comparison with experimental data. Phys. B Condens. Matter. **484**, 27–30 (2016). https://doi.org/10.1016/J.PHYSB.2015.12.046
4. Dwara, S.N., Al-Khursan, A.H.: Quantum efficiency of InSbBi quantum dot photodetector. Appl. Opt. **54**, 9722 (2015). https://doi.org/10.1364/ao.54.009722
5. Mal, I., Panda, D.P., Tongbram, B., Samajdar, D.P., Chakrabarti, S.: Analytical modeling of temperature and power dependent photoluminescence (PL) spectra of InAs/GaAs quantum dots. J. Appl. Phys. **124**, 145701 (2018). https://doi.org/10.1063/1.5047026
6. Mal, I., Samajdar, D.P., John Peter, A.: Theoretical studies on band structure and optical gain of GaInAsN/GaAs/GaAs cylindrical quantum dot. Superlattices Microstruct. **119**, 103–113 (2018). https://doi.org/10.1016/J.SPMI.2018.04.043
7. Mal, I., Samajdar, D.P., Das, T.D.: Effect of Sb and N resonant states on the band structure and carrier effective masses of GaAs1-x-yNxSby alloys and GaAs1-x-yNxSby GaAs quantum wells calculated using k·p Hamiltonian. Superlattices Microstruct. **106**, 20–32 (2017). https://doi.org/10.1016/j.spmi.2017.03.035
8. Mal, I., Samajdar, D.P., Das, T.D.: Calculation of band structure and optical gain of type-II GaSbBi/GaAs quantum wells using 14-band k·p Hamiltonian. Superlattices Microstruct. **109**, 442–453 (2017). https://doi.org/10.1016/J.SPMI.2017.05.032
9. Mal, I., Jayarubi, J., Das, S., Sharma, A.S., Peter, A.J., Samajdar, D.P.: Hydrostatic pressure dependent optoelectronic properties of InGaAsN/GaAs spherical quantum dots for laser diode applications. Phys. status solidi. (2018). https://doi.org/10.1002/pssb.201800395
10. Mal, I., Samajdar, D.P.: InSbNBi/InSb heterostructures for long wavelength infrared photodetector applications: A 16 band k·p study. J. Appl. Phys. **128** (2020). https://doi.org/10.1063/5.0020813
11. Vurgaftman, I., Meyer, J.R., Ram-Mohan, L.R.: Band parameters for III-V compound semiconductors and their alloys. J. Appl. Phys. **89**, 5815–5875 (2001). https://doi.org/10.1063/1.1368156